A Handbook for Clinical Investigators Conducting Therapeutic Clinical Trials Supported by CTEP, DCTD, NCI

Table of Contents

Table of Contents

Table of Contents

Table of Contents

Table of Contents

This handbook explains the policies and implementing procedures for the conduct of therapeutic clinical trials sponsored by the Division of Cancer Treatment and Diagnosis (DCTD), National Cancer Institute (NCI). Sponsorship or support of clinical trials includes funding, regulatory support and/or agent distribution. Oncologists, nurses, pharmacists, research administrators, and data managers should find the information in this handbook useful in practical matters connected with protocol drafting and submissions, reporting requirements, agent accountability, and a host of other subjects. This handbook is written to guide the individual clinical investigator at the clinical trial site working alongside a team of health professionals and research staff. Clinical studies are often conducted as a multi-center or Cooperative Group trial where the Coordinating Center or Group Operations assumes responsibilities for various functions. Although specific terminology may differ among these trial organizations, the principles and responsibilities discussed in this handbook apply to all investigators and participating institutions in therapeutic clinical trials supported by CTEP, DCTD, NCI.

The Cancer Therapy Evaluation Program (CTEP) of the DCTD (http://dctd.cancer.gov) is responsible for implementing and monitoring the clinical development of investigational therapeutic anticancer agents. CTEP's policies are intended to ensure patient safety while providing the National Cancer Program with the most effective new agent development program possible. Some policies reflect the regulatory requirements of the Food and Drug Administration (FDA) and the Department of Health and Human Services (DHHS). Others have been developed based on policies at the Institute level, consensus among CTEP staff, the NCI Board of Scientific Advisors, and leaders in the community of clinical investigators. Specific policies and procedures continue to evolve; through them CTEP, DCTD, NCI aims to provide a flexible, responsive system within the constraints imposed by regulation and the program's size and scope.

In addition to the scientific and medical issues involved in planning and conducting clinical research, CTEP has two additional major responsibilities:
>   (1) Sponsorship of Investigational New Drug Applications (INDs) and
>   (2) Oversight of the contracts, grants and cooperative agreements under which most clinical testing takes place.

In this handbook we refer to DCTD as the IND sponsor and the "proprietor" of the investigational agent development program and to CTEP where the efforts of the CTEP staff are specifically involved. The use of NCI is reserved for more general contexts, including overall support of clinical trials.

The CTEP web site (http://ctep.cancer.gov) is an excellent additional resource for current information on CTEP policies, forms/templates and initiatives. Familiarity with CTEP's organization and the primary responsibilities of each branch or office may help you understand how CTEP functions.

We welcome readers' comments on this handbook's content and how future updates can make this handbook more useful. Please contact NCICTEPiHandbook@mail.nih.gov.

Jeffrey Abrams, M.D., Associate Director, Cancer Therapy Evaluation Program
Jan M. Casadei, Ph.D., Chief, Regulatory Affairs Branch, CTEP

Investigator Handbook Committee

Julie K. Rhie, Ph.D., R.Ph., Chair	Joan Mauer
Jeannette Wick, R.Ph., MBA, Editor	Margaret Mooney, M.D., M.B.A.
Sherry Ansher, Ph.D.	Larry Rubenstein, Ph.D.
Austin Doyle, M.D.	Patricia R. Schettino, R.Ph., M.S.
Shanda Finnigan, R.N.	Gary Smith, M.T., M.G.A.
Charles Hall, M.S., R.Ph.	Naoko Takebe, M.D., Ph.D.
Martha Kruhm, M.S.	Jian Zhang, Ph.D.

When clinical investigators conduct clinical studies two elements are crucial – the trial's sponsor and the clinical trial site. Sections 2 and 3 discuss the purposes and features of each.

In Section 2 we describe DCTD's role as a sponsor of investigational agent trials. Specifically, we review CTEP's responsibility for overall direction of the process of investigational agent development and its practical implementation. We outline the basis of the relationship between DCTD, pharmaceutical collaborators, and investigators during the conduct of clinical trials.

2.1   The Sponsor

Development of oncology agents is a long and complex process, but successes have been significant. The fact that some patients with aggressive neoplasms now have long term survival is the best possible evidence that agents with selectivity against cancer can be identified and used effectively. On the other hand, the oncology community is well aware that for many tumor types, systemic treatment is unsatisfactory. The motivation to develop better therapy is therefore as powerful as ever. With the increased understanding of the malignant process due to recent and anticipated advances in molecular biology, cancer genomics and biochemical pharmacology, we have every reason to expect that the development of new agents will proceed along increasingly rational lines.

The process of new agent development is often divided into preclinical and clinical components. Although this division is operationally useful, continual interplay exists between these arenas. Evidence of synergy or the effectiveness of combined modality approaches in experimental models, for example, has provided the major motivation for a large number of clinical trials. The converse is also true; clinical observations have also given rise to new lines of basic investigation.

Historically, the NCI has been one of the most important effectors in the discovery and development of new anticancer agents. NCI's prominent role in new cancer agent development has no parallel elsewhere in developmental pharmacology. The justification for such intensive involvement of a Government agency in research and development is clear: significant improvement in cancer treatment is in the public interest. NCI is the largest clinical trials sponsor focused on cancer treatment and diagnosis, and currently has a significant number of new agents in various stages of clinical testing or preclinical development.

As part of this massive effort, NCI funds a clinical trials network that includes Cooperative Groups and Consortia, new agent development contractors, investigators at Cancer Centers, University hospitals and Specialized Programs of Research Excellence (SPOREs). More than 20,000 investigators from approximately 3,000 institutions participate in this effort.

In the United States, clinical research with investigational agents is carefully regulated. The regulatory authority for assuring public safety in matters relating to investigational drugs and biologics rests with the Food and Drug Administration (FDA, http://www.fda.gov/). FDA regulations, which are specific implementations of the Food,

Drug, and Cosmetic Act and the Public Health Service Act define the terms under which clinical work with investigational drugs and biologics may proceed (see 21 CFR 312 and 21 CFR 600). Because these regulations have the force of law, all those involved in clinical trials with investigational agents must heed these laws, including NCI, pharmaceutical collaborators, and investigators. An organization or an individual that assumes these legal responsibilities for supervising or overseeing clinical trials with investigational agents is termed an IND sponsor. In the United States, the DCTD and pharmaceutical collaborators most commonly sponsor such research in cancer. The designation obviously implies a substantial commitment of resources.

In addition, the Public Health Service Act mandates a number of safeguards for the rights and welfare of individuals who are involved as research subjects. Department of Health and Human Services (DHHS) regulations, administered by the Office for Human Research Protections (OHRP), DHHS, specify requirements in addition to those of the FDA to ensure adequate human subject protection. Clinical investigators and institutions taking part in the clinical trials network are responsible for meeting the requirements of the HHS regulations. In addition, institutions conducting clinical trials must also abide by the Health Insurance Portability and Accountability Act (HIPAA).

As a sponsor of investigational agents, DCTD, and specifically CTEP, is responsible for seeing that clinical trials proceed safely and rationally from the initial dose-finding studies to a definitive evaluation of the role of the new agent in the treatment of one or more specific cancers. Fulfillment of this goal obviously requires active participation of DCTD staff throughout the entire process.

2.2   How NCI Funds Research

A full discussion of the means by which NCI funds research is beyond the scope of this handbook. Whether support comes from investigator-initiated grant, contract, or cooperative agreement, however, the peer review process is central. Government officials can provide monies to investigators only in the context of mechanisms involving peer review; this process requires formal application by the investigator and (usually) multiple levels of evaluation. Once an application is approved, the NCI cannot provide more funding than is stipulated by the judgment of peer review and the NCI Board of Scientific Advisors. Additional awards can, of course, be made after review and formal approval of a supplemental application. Provision of investigational agents is separate from clinical study funding. However, CTEP will make a good faith attempt to supply investigational agents required for funded research proposals.

2.3   Preclinical Development of Investigational Agents

The DCTD accomplishes its overall aims in investigational agent development by building on its extensive preclinical efforts. The DCTD Developmental Therapeutics Program (DTP, http://dctd.cancer.gov/ProgramPages/dtp/default.htm) is committed to the discovery and development of new anticancer agents. Please refer to a book chapter by Boyd, MR in Status of the NCI Preclinical Antitumor Drug Discovery Screen, *Principles and Practice of Oncology Updates,* Vol. 3(10): 1-12 (1989) at http://home.ncifcrf.gov/mtdp/Catalog/full text/Paper309/Paper309.pdf for a summary of methods utilized by DTP for preclinical agent discovery and development. The mission of the DCTD Chemical Biology Consortium (CBC) (http://dctd.cancer.gov/CurrentResearch/ChemicalBioConsortium.htm) is to increase the flow of early stage drug candidates into NCI's drug development pipeline. Leads from this program will flow into the NCI's Experimental Therapeutics (NExT) Program. The (NExT) Program (http://next.cancer.gov/about/default.htm) is a partnership between NCI's Division of Cancer Treatment and Diagnosis (DCTD) and the Center for Cancer Research (CCR). Agents accepted into the NExT Program are eligible for developmental resources from drug discovery to early phase 2 clinical trials. SPOREs also support investigator-initiated early phase clinical trials and relevant pre-clinical studies. More about SPOREs can be found at http://trp.cancer.gov/.

2.4   Collaboration between DCTD and the Pharmaceutical Collaborators

Many of the anticancer agents in CTEP's pipeline result from a co-development venture with a pharmaceutical collaborator. Collaboration between DCTD and the pharmaceutical industry may occur at any step along the new agent development process. Private companies often submit agents to DCTD for testing and joint development. Agents may be submitted for antitumor screening, preclinical toxicology, or clinical testing. Conversely, if DCTD discovers an agent, a pharmaceutical collaborator is sought early in development, since DCTD does not market new agents. Early pharmaceutical collaborator involvement permits substantial cost-sharing between public and private sectors, and can hasten the availability of effective agents by several years. Please refer to: http://ctep.cancer.gov/industryCollaborations2/default.htm#guidelines for collaborations

Development plans for new agents, therefore, are usually a collaborative effort between DCTD and the pharmaceutical collaborator. The DCTD and pharmaceutical collaborator usually formalize their collaboration with a Cooperative Research and Development Agreement (CRADA) or Clinical Trials Agreement (CTA). The model CRADA and its appendices can be found at http://ctep.cancer.gov/industryCollaborations2/model agreements.htm. In this joint effort, DCTD and the private sector have a common goal: defining a new agent's contribution to cancer treatment as precisely and expeditiously as possible. Clinical investigators' opinions are often sought while formulating development plans for an agent. The timely approval of a new agent by a New Drug Application (NDA) or a Biologic License Application (BLA) by the FDA is in the public's interest. However, there may well be differences between these partners in sponsoring certain kinds of trials. The complex three-way relationship among clinical investigators, the DCTD, and private industry means coordinating efforts, establishing priorities, and allocating limited resources. To facilitate interactions, CTEP has developed guidelines on the nature of the relationship between the participants. The guidelines, which formally recognize the private sector's involvement in and support of clinical trials, are summarized in "NCI-Cooperative Group-Industry Relationship Guidelines" (See Appendix I), http://ctep.cancer.gov/industryCollaborations2/guidelines.htm

2.5   Private Support of Trials Supported by NCI Funding

As private support for clinical trials in cancer becomes more widespread, investigators and Cooperative Groups holding grants, contracts, or cooperative agreements from NCI should carefully consider the allowable allocation of resources provided by a private entity for a trial already receiving NCI support. Investigators and Cooperative Groups must make certain that Federal funds are not used to cover those costs of research also supported by private resources. Grants management personnel at NIH and auditors from DHHS are required to scrutinize such arrangements closely and may take steps to recover Federal funds if they have been used inappropriately.

In the specific case of the clinical Cooperative Groups, the Terms of Award of NCI's current Cooperative Agreements permits them to accept industrial support, provided that industry funds are limited to costs not funded by the NCI and they are appropriately accounted for and reported.

In the case of NCI-funded phase 1 and phase 2/3 trials, private provision of resources for tasks not supported by Federal funds may also be appropriate; investigators should discuss all such requests with CTEP.

2.6   Private Support of Trials

Private support of a trial sponsored under a DCTD-held IND is appropriate under certain circumstances. However, the Protocol/Group Chair and the private firm should draft a written agreement, and send a copy of the draft agreement to the Associate Chief, Regulatory Affairs Branch (RAB), CTEP for review and approval. In general, CTEP will favor the provision of data from trials of this kind to a pharmaceutical collaborator. The investigator's obligations to DCTD as the sponsor, as detailed throughout this handbook, remain unchanged.

2.7   The Investigational New Drug Application (IND)

Any organization or individual investigator seeking to sponsor clinical trials with investigational agents must first submit an IND to the FDA. The use of the term "sponsor" is generally reserved for organizations assuming broad responsibilities for the development of a new agent. It is also possible for individual investigators to hold an IND (see 21 CFR 312.6, "sponsor-investigator"). The IND is the legal mechanism under which investigational agent research is performed in the United States. No investigational agents may be administered to patients for research in the U.S. without an active IND.

The FDA's regulations for drugs and biologics found in 21 CFR 312 and 21 CFR 600, respectively, (http://www.accessdata.fda.gov/scripts/cdrh/cfdocs/cfcfr/CFRSearch.cfm) specify an IND sponsors' obligations. The DCTD, while a component of an agency in the DHHS, is just as accountable, e.g. as a pharmaceutical company for meeting these obligations.

The sponsor's initial IND submission to the FDA is a lengthy document that sets forth the experimental rationale for human testing, based on results of animal pharmacology and toxicology studies, manufacturing data, purity and stability information, and provides an initial plan of clinical investigation.

The IND is the official record at the FDA of the sponsor's clinical research with the agent. Under FDA regulation, DCTD must maintain the IND as an accurate, timely repository of all information concerning the agent's clinical use, including all clinical protocols, clinical protocol amendments, adverse events, and an annual report on all clinical trials and any new relevant preclinical (particularly toxicological) or manufacturing data. Obviously this means that no one can use an investigational agent without the IND sponsor's knowledge and prior approval.

After a sponsor submits an IND, FDA has 30 days to complete its review. If FDA has safety concerns, it may place a hold on the initiation of a clinical trial(s) with the agent. Please note that these are matters between the sponsor and the FDA. An investigator may not initiate patient treatment on a protocol using DCTD-sponsored agents until he or she has received written notice of approval from CTEP.

2.8   The Marketing Application

After clinical trials have shown that the new agent is safe and effective, the formal process by which an agent is made available to the public in the U.S. is FDA approval of a marketing application (New Drug Application or a Biologic License Application). These are submitted by a pharmaceutical company; as noted previously, NCI does not submit NDAs or BLAs since it does not market products. The applicant seeks approval from FDA for one or more specific indication(s). Review and approval of an NDA or BLA are based on demonstration of safety and efficacy assessed from detailed reports of the clinical trials, particularly randomized controlled studies. A new agent's contribution to the treatment of a disease is demonstrated unambiguously if the agent is the only variable between the treatments.

The specific endpoints that constitute satisfactory evidence of efficacy (e.g., response rate, quality of life, and survival) have been addressed in a published paper prepared by FDA and NCI entitled "Commentary Concerning Demonstration of Safety and Efficacy of Investigational Anticancer Agents in Clinical Trials." This paper was prepared with input and advice from the Oncologic Drugs Advisory Committee of FDA (a panel of outside experts in clinical oncology) and the Board of Scientific Counselors of the DCTD (a panel of outside experts in preclinical and clinical oncology). A copy of this paper is available at http://jco.ascopubs.org/cgi/content/abstract/9/12/2225.

The approval of the NDA or BLA is a critical milestone not only for the pharmaceutical company but also for the clinical investigator, the practicing oncologist, NCI, and the general public. An affirmative decision by the FDA permits the pharmaceutical company to market and promote the agent for the approved indication(s). Once an agent is marketed, no Federal regulation prevents licensed physicians from prescribing it for any indication they deem appropriate.

For the NCI, NDA or BLA approval marks a step forward in the development of effective cancer therapies. Following approval the NCI usually continues to sponsor further research with the agent for other indications or disease settings.

In Section 3 we discuss the clinical trial site's vital role in support of the clinical investigator. As is the case with all types of research, clinical trials require a substantial institutional commitment.

3.1   Definition and Purpose of a Clinical Trial Site

A clinical trial site is an entity/institution that assumes a broad range of responsibilities and functions for the support of clinical trials conducted under its name. Examples of a clinical trial site include a single institution (i.e., academic, medical center, hospital, clinic), multi-center study participants, NCI-designated Cancer Centers (refer to http://cancercenters.cancer.gov), Cooperative Groups, Consortia, and SPOREs. The clinical trial site supports investigators as they develop, organize, implement and analyze clinical trials.

Clinical studies are often conducted as a multi-center or Cooperative Group trial where the Coordinating Center or Group Operations assumes the responsibilities for various functions. Although specific terminology may differ among these trial organizations, the principles and responsibilities discussed in this handbook apply to all clinical investigators and clinical sites participating in therapeutic clinical trials supported by CTEP, DCTD, NCI. The clinical trial site assumes responsibility for research quality, both in concept and execution, and ensures patient safety.

An effective clinical trial site enhances the investigator's research in several specific ways. It provides assistance in developing protocols and obtaining approval by sponsoring agencies. It often offers centralized data management and statistical consultation. An effective clinical trial site should also provide the opportunity for internal peer review and quality assurance.

The clinical trial site enhances its own scientific credibility by assuming responsibility for the quality of the scientific ideas and the care with which they are tested. These activities may also be an economical way of supporting multiple clinical investigations simultaneously.

3.2   Activities of a Clinical Trial Site

3.2.1   Protocol Development-Scientific Review and Biostatistical Consultation

Many clinical trial sites have internal procedures for review of each clinical trial's science, either at the concept stage or at the time a protocol is written. These reviews are distinct from the task of the Institutional Review Board (IRB), which are directed at patient protection and may or may not provide critical scientific review. Ideally, scientific review helps investigators focus their ideas. It may also help identify other useful scientific resources within the clinical trial site. As a whole this process should facilitate research and help test new ideas. Additional scientific and statistical review should be conducted for phase 3 studies, due to the substantial commitment of time, patients, and resources involved.

All clinical trial designs should be based on sound statistical principles. Issues such as sample size, stopping rules, endpoints, and the feasibility of relating endpoints to objectives are pivotal to a successful trial. Statistical review should be provided by experts at the clinical trial site.

3.2.2   Protocol Administration

Most protocols require multiple levels of approval. Since policies may change with time; clinical trial sites should assist investigators and ensure they obtain these approvals. Establishment of a centralized mechanism for submitting and tracking a protocol through the necessary approvals, including the IRB, saves individual investigators time and effort that is better directed elsewhere. A clinical trial site engaged in CTEP-supported protocols must assume responsibility for communicating status changes, amendments, results reports, publications, and other pertinent protocol administration information to CTEP.

3.2.3   Establishment of an Affiliate Program

An affiliate investigator is a physician who participates in a clinical trial organized by a lead institution and has satisfied all criteria for affiliate membership as defined by the primary institution (see Section 13). Engaging affiliate investigators often contributes to cancer clinical trial success. The clinical trial site must be as concerned about the quality of research performed by its affiliate investigators as it is with that of its own staff. The Principal Investigator must oversee affiliate investigators closely.

CTEP has established a set of guidelines to assist lead clinical trial sites in developing a policy toward affiliate investigators (see Section 13). Each clinical trial site participating in study agent trials supported by CTEP and/or sponsored by DCTD may develop its own affiliate policy in accord with these guidelines, provided they meet all CTEP Guidelines.

The most important components of these guidelines are that the lead clinical trial site should, (a) define qualifications necessary for affiliate investigators and (b) periodically review their performance. Such review of performance should include site visits by investigators from the lead clinical trial site.

3.2.4   Agent Accountability and Storage

Although FDA regulation holds the investigator accountable for the proper use of investigational agents, designated staff may assume these responsibilities for investigators. Regardless, a study's Principal Investigator is ultimately responsible for compliance with Federal requirements (see Section 15).

3.2.5   Reporting of Results to CTEP

Investigators on CTEP-supported studies (or organizations that assume responsibility for investigators) must provide the following to CTEP:
- adverse events
- protocol amendments
- protocol status change
- periodic study results and final study results
- publications

The summaries and status of all CTEP studies must be reported via the mechanism best suited for the study. Section 10 provides additional detail on reporting guidelines. The

Cooperative Groups and Consortia inform CTEP of this information directly. In all cases, the Protocol Chair or Group Chair is responsible for reporting to CTEP.

3.2.6   Data Management and Statistics

Since most cancer clinical trials involve professional staff other than the Protocol Chair or Group Chair, the institution must integrate proper collection of clinical data into their medical practices. Data collection is best done as data are generated; this practice promotes protocol compliance and permits the Protocol Chair/Group Chair to monitor the study's progress. For these reasons, data management organized and supported at the department or institution level is usually more efficient and reliable than that which is left to the individual investigator. CTEP does not require centralized data management by institutions performing CTEP-supported trials, but this approach is highly recommended. In the experience of CTEP's site visit monitoring program (see Section 16), institutions that provide central support for data management tend to have better quality than those without.

3.2.7   Quality Assurance

Clinical trial investigators are obligated to take appropriate steps to protect the integrity of science and the safety of human subjects (study participants). Selecting responsible co-investigators and research staff is the best way to protect against protocol deviations or poor data quality.

Quality assurance includes prevention, detection, and action from the beginning of data collection through publication of the results. Special efforts should be made to assure protocol adherence, adequate assessment of eligibility, compliance with protocol treatment and regulatory requirements, timely adverse event reporting and complete collection of data on the primary outcome measures. Another goal of a quality assurance program is to detect problems by implementing routine monitoring procedures and periodic data audits.

These activities take many forms. Section 16 describes existing CTEP and Cooperative Group procedures for quality assurance and protocol deviations. CTEP encourages even single institutions to establish internal quality assurance programs. CTEP staff will assist and provide advice to any clinical trial site that wants to develop these programs.

3.2.8   Biomarker and Imaging Studies

Clinical trials now often use laboratory assays or imaging modalities to assess or investigate novel biomarkers. These investigational biomarkers are specific to the disease or the agent (as opposed to routine measurements of normal organ function such as CBC, serum albumin, etc.), but they are not yet part of the generally accepted diagnostic algorithm for the disease. Exceptions include the relatively few biomarkers that are currently accepted, such as the ER/PR and HER2 status of breast cancers. A biomarker may be integral to the trial, or it may be included as either an integrated or correlative study. It is also clear that as more targeted agents are evaluated in clinical trials, the quality of the assay to measure a molecular or imaging marker that is either a direct or indirect target of the agent must improve in order to quickly and efficiently assess the agent's safety and efficacy. As described below, this is especially critical for markers that are essential or integral for late phase trials. However, pharmacodynamic markers in early phase trials may also need to meet similar criteria for analytical performance for appropriate early analysis of agent efficacy.

An *integral* biomarker is a test that must be performed on all subjects in real time for the trial to proceed, such as one that is used to establish eligibility, to assign treatment or to stratify the case for randomization. For integral biomarkers, laboratory or imaging tests must meet performance standards suitable for clinical practice (see Performance Standards Reporting Requirements for Essential Assays in Clinical Trials, http://cdp.cancer.gov/scientificPrograms/pacct/assay_standards.htm). The investigator(s) responsible for biomarker measurements must be identified. By law, investigators must employ a CLIA-certified laboratory to perform any laboratory test if the result will be used to assign treatment, or will be reported to the patient or his/her physician at any time; additional state or local regulations may apply. Information about the integral biomarker should be included in the protocol in the sections for Objective(s), Background and Rationale and Statistical Considerations, and in other sections of the document as appropriate (e.g., Patient Eligibility Criteria). This information should also include a brief description of the analytical performance characteristics of the assay for the biomarker. Guidelines may be found in Table 2 of Dancey *et. al.* (see Guidelines for the development and incorporation of biomarker studies in early clinical trials of novel agents. Clin Cancer Res. 2010 Mar 15;16(6):1745-55 at http://www.ncbi.nlm.nih.gov/pubmed/20215558). Trial registration procedures and accrual targets must consider anticipated turn-around times for test results and expected rates of test failure. It is essential that the protocol include clear and complete instructions for the acquisition, processing and shipment of specimens and/or imaging data. For additional information please refer to the Cancer Diagnosis Program website at http://www.cancerdiagnosis.nci.nih.gov.

Integrated biomarkers are not used to determine treatment in the trial, but are clearly identified as part of the trial from the beginning and are intended to identify or validate tests planned for use in future trials. Integrated tests are performed on all subjects or on a pre-specified subset of cases, and the study includes complete plans for specimen or image collection, laboratory measurements and statistical analysis. The investigator(s) responsible for biomarker measurements must be identified. Even if laboratory test results are not used to determine treatment, if the results will be reported to the patient or his/her physician at any time, then the test must be performed in a CLIA-certified laboratory; additional state or local regulations may apply. As for integral biomarkers, information about the test should be included in the protocol's Objective(s), Background and Rationale and Statistical Considerations sections, and in other sections as appropriate (e.g., Special Studies or an Appendix). It is essential that the protocol document itself include clear and complete instructions for the acquisition, processing and shipment of specimens and/or imaging data; if specimen collection is mentioned in more than one section of the protocol (for example, in both Procedures for Patient Entry on Study and in an Appendix), be careful to maintain consistency in the instructions.

Correlative or ancillary laboratory or imaging studies may be retrospective or exploratory. To enhance opportunities for high-quality retrospective studies, which can be exceptionally informative, investigators are encouraged to build plans for specimen collection and tissue banking into clinical trials. Correlative studies of tests that do not meet the criteria above for either integral or integrated biomarkers may be embedded in the protocol. See Guidelines for Correlative Studies in Clinical Trials (http://ctep.cancer.gov/protocolDevelopment/default.htm#ancillary_correlatives) for information that should be included in the protocol for embedded correlative studies.

3.2.9   Unblinding Procedure for Placebo-Controlled (Blinded) Studies

The blind *may* be broken for a serious and unexpected event, if it is essential for the medical management of the subject, *or* may provide critical safety information about a drug that could have implications for the ongoing conduct of the trial (e.g., monitoring, informed consent). In those rare instances where it is considered necessary, steps must be in place to reveal the treatment assignment of the patient in question. To that end, CTEP requires that unblinding procedures be included in the agent information section of all placebo-controlled protocols. The protocol should contain sufficient contacts to ensure unblinding is possible 24 hours per day, 7 days a week.

These unblinding procedures should include: who at the site level is authorized to request the unblinding, what justification to unblind is acceptable, the contact number of the person authorized to approve the unblinding, and who may provide the blinded treatment assignment. The section should also state the subject's role in the study once the treatment assignment is revealed.

The Development of a Clinical Trial

The following section explains the policies for phase 1 clinical investigations with CTEP study agents. We outline the scientific objectives for CTEP-supported clinical trials and describe which physicians are eligible to study and administer study agents. These two aspects of agent use-study and administration-are formally separate issues. In general, eligibility to study CTEP study agents is restricted to institutions and investigators approved by peer review; these individuals may include grantees, contractors, Cooperative Group members, physicians affiliated with approved cancer centers, and recipients of investigator-initiated clinical research project grants (RO1, PO1). Except in certain explicitly defined circumstances outlined in the following sections, the administration of study agents is restricted to these investigators. Information regarding the use of investigational agents in phase 0 studies is found in Appendix II.

For investigational agents involving a pharmaceutical collaborator and a DCTD-sponsored IND, the Investigational Drug Steering Committee (IDSC) is consulted to develop a strategic clinical development plan. The IDSC membership includes Principal Investigators from NCI Phase 1 U01 grants and Phase 2 N01 contracts, as well as representatives from Cooperative Groups. Its members include experts in biostatistics, imaging, radiation oncology, and clinical and preclinical pharmacology. The IDSC subgroups assist CTEP with early phase trial prioritization, and address critical scientific issues for the trials
(http://ctep.cancer.gov/SpotlightOn/Investigational Drug Steering Committee.htm).

4.1   Scientific Policies of CTEP

4.1.1   Planning of Phase 1 Trials

CTEP plans the phase 1 development of each agent prospectively, basing schedule selection and starting doses for phase 1 clinical trials on experimental data. Generally, each schedule is examined in not more than two studies.

From the results of the phase 1 and clinical pharmacology studies, CTEP and its collaborators (investigators, pharmaceutical collaborators and IDSC) decide which schedule to bring forward into phase 2. An additional schedule would initially undergo a safety evaluation in a phase 1 trial. Should there be a desire to compare the two schedules for efficacy that comparison would be done in a phase 2 trial.

4.1.2   Objectives

Phase 1 trials determine a safe and/or biologically effective dose for phase 2 trials and help define adverse effects on normal organ function. In addition, these trials examine the agent's pharmacology which may reveal evidence of antitumor activity. Therapeutic intent is always present in Phase 1 trials; indeed, anticancer agents are not tested in patients unless preclinical activity studies have demonstrated evidence of significant activity in laboratory animals or an appropriate *in vitro* model.
Animal toxicology studies conducted prior to phase 1 trials provide the investigator with
- recommended starting dose for clinical trials
- agent's anticipated effects (desired and adverse) on normal tissues

These studies provide investigators with information that helps focus clinical observation of the patient. The phase 1 dose is increased gradually by defined procedures until a level is found that

- produces limiting but tolerable adverse events, or
- finds pharmacodynamic (PD) evidence of target effect, and/or
- finds clear signs of therapeutic activity.

Phase 1 studies may have multiple endpoints, including determination of a biologically effective dose, but usually increase doses to some level of tolerable toxicity. Phase 1 trials define acute effects that occur with a relatively high frequency. Continued careful observation during phase 2 and 3 trials is essential to identify less frequent acute adverse effects, as well as cumulative and chronic adverse events.

4.1.3   Patient Selection

Patients eligible for phase 1 trials must have confirmed malignant disease that is not effectively treated by conventional forms of therapy or for which there is no standard treatment. Initially, patients should have normal organ function so the investigator may reliably distinguish agent effects from disease effects.

In the presence of major organ impairment, drug treatment may increase adverse effects due to decreased clearance or additive injury to the organ. Since most anticancer agents will ultimately be used in some patients having impairment of major organ function (particularly cardiac, hepatic, and renal), it is reasonable to explore their use in these populations. CTEP sponsors specific phase 1 trials explicitly designed to determine safe and tolerable doses in patients who have organ impairment and to determine potential toxicities (see Section 4.1.9). CTEP usually supports such trials selectively after the initial trials in patients with normal organ function.

After successful completion of phase 1 trials in adults with normal and abnormal physiology, CTEP initiates other types of studies. These studies include those in pediatric populations and elderly patients. Such attempts are aimed at identifying a range of maximally tolerated doses (MTDs) or biologically effective dose for each agent.

Typically, phase 1 studies are performed in both women and men. If gender-specific pharmacologic differences exist, these differences must be demonstrated and characterized. In some situations, pharmacogenomic differences in drug metabolism or target effects may alter toxicity or efficacy of the agent in different ethnic populations, requiring separate analyses of drug effects in these ethnic groups. Additionally, disease-specific phase 1 studies may occasionally be conducted particularly in the setting of trials involving a combination of agents.

4.1.4   Schedule Selection

Several factors determine the number of separate schedules studied in phase 1 trials. These include evidence of schedule dependence in experimental *in vivo* systems; pharmacokinetics, mechanism of action, if known; and existing clinical data with similar compounds suggesting superiority of a particular schedule. Agents that are highly schedule-dependent in preclinical models are usually brought into phase 1 trials on the putative optimal schedule.

Because the correlation between schedule dependence in preclinical models and the clinic is not firmly established, however, some agents may be candidates for a broader array of schedules. DCTD is prospectively evaluating the ability of experimental models to predict the schedule dependence of activity, adverse effects, and pharmacokinetics. For agents showing no particular schedule dependence in preclinical models, two extremes of schedules (e.g., single bolus dose per course and 5-day continuous infusion) are sometimes examined.

4.1.5   Starting Dose

The starting dose of a phase 1 trial is based on preclinical toxicology studies. Numerous formulas to calculate a starting dose exist, for example, a fraction of the MTD in the most sensitive species tested or the no adverse effect level (NOAEL). Investigators would discuss their rationale for the proposed starting dose with CTEP.

4.1.6   Dose Escalation

Doses are generally escalated according to a schema in which incremental increases in dose decrease as biologic activity becomes evident. Often, a modified Fibonacci plan is employed. However, when the goal is to escalate to a biologically effective dose as rapidly as possible, a number of accelerated titration, continual reassessment method, and other designs allow rapid dose escalation in the absence of limiting toxicity. A frequent phase I design employs successive dose doubling until a Grade 3 adverse event or two instances of a Grade 2 adverse event are seen. The exact schema may be affected by the steepness of the dose toxicity curve in animal models or, for trials of combinations of agents, the steepness of the single-agent dose toxicity curves. In all cases, the goal is to arrive at the recommended phase 2 dose with the fewest number of escalations consistent with patient safety; this approach minimizes the number of patients receiving biologically inactive doses. CTEP is actively evaluating other methods of dose escalation, based on the use of pharmacodynamic markers to determine target effects, and the accelerated titration designs for phase 1 clinical trials found at http://linus.nci.nih.gov/~brb/Methodologic.htm.

When there is sufficient concern about anticipated adverse events, a minimum of three patients not previously treated with the new agent should be entered at each dose level. In these cases, escalation to the next level should not occur until the safety of the current level has been established. This may require observation of at least three patients for the entire course interval (e.g., 3 – 5 weeks). For many trials, however, escalation can proceed with one or two patients per level provided no Grade 3 or repeated Grade 2 adverse events have yet been seen in the study. Intra-patient dose escalation should be considered for use wherever it is deemed safe. At least six patients should be treated at the recommended phase 2 dose. The incidence of dose limiting toxicity acceptable for a recommended dose should be specified in the protocol (e.g., <33%).

4.1.7   Pharmacokinetics

Pharmacokinetic determination of drug levels and metabolism of parent drug and principal metabolites during phase I studies can confirm optimal schedules for drug administration; determine the principal routes of human metabolism; predict populations at particular risk for drug toxicity; and suggest whether flat dosing or BSA-based dosing of drug is appropriate for a particular agent. The role of pharmacokinetics in phase 1 trials is now receiving increasing emphasis, with specific focus on the possible use of such data to guide dose escalation. Investigators developing phase 1 trials should consider pharmacokinetic determinations an integral part of their study. Investigators with questions regarding suitable PK methodologies for their trials should check with the Investigational Drug Branch (IDB) staff physician before writing a protocol.

4.1.8   Studies to Determine Pharmacodynamic Effects

Pharmacodynamic (PD) analysis of the biochemical and physiologic effects of anticancer drugs on the body, and on specific drug targets in the cancer cell, is an increasingly important component of early phase cancer trials. Rational, efficient development of endpoints of drug effect on target, pathway, and downstream biological processes can:

- lead to more rapid and efficient development of targeted cancer therapeutics,
- help determine the biologic effective dose of the agent.

Determination of PD endpoints can be linked to the therapeutic effects of the investigational agent and help provide proof of concept for target modulation. PD biomarkers in early clinical trials inform the rational selection of the agent's dose and schedule, and may explain or predict clinical outcomes. (Adv Cancer Res. 2007;96:213-68. at http://www.ncbi.nlm.nih.gov/pubmed/17161682 and Clin Cancer Res. 2010 Mar 15;16(6):1745-55 http://www.ncbi.nlm.nih.gov/pubmed/20215558)

4.1.9   Imaging Studies to Determine Pharmacodynamic Effects

Increasingly, imaging technology is able to perform functional or molecular imaging of cancers. These modalities visualize physiological, cellular, or molecular processes in tumors. They allow observation of anticancer agents' effects on tumors, permitting treatment monitoring and providing evidence of target effects. DCTD's Cancer Imaging Program (CIP) is an innovative biomedical program to advance understanding of cancer imaging. CIP supports and advises innovative imaging correlates to early phase clinical trials by funding projects in key areas and developing protocols for imaging studies of anticancer drug effects (http://dctd.cancer.gov/ProgramPages/cip/default.htm). CTEP and CIP often collaborate on clinical studies that combine an investigational study agent and an imaging component.

4.1.10 Phase 1b Studies

CTEP supports phase 1b studies which are studies of the initial combination of investigational agents with standard anticancer agents or with other targeted therapies. These combination trials are often started after initial trials of the investigational agent have shown evidence of tolerability and some evidence of single-agent activity. Single agent studies may be unnecessary in situations where the new investigational agent acts as a cytotoxicity potentiator, lacking expected inherent anticancer activity. Investigators should provide a strong rationale for the drug combination, supported by preclinical studies, for proposed trials. IND agents may be combined if there is sound evidence to support this study design. These combination investigational agent studies are evaluated on a case-by-case basis.

4.1.11 Phase 1 Organ Dysfunction Trials

Special populations are generally excluded from studies of investigational agents because dosing or scheduling information is unknown or patients are considered too frail to tolerate treatment. Many oncology drugs are approved by the FDA with limited PK/PD information in patients with organ dysfunction. Cancer patients who have renal or hepatic organ dysfunction may require dose reductions or modifications. The number of cancer patients with impaired hepatic or renal function eligible for protocols specifically

evaluating organ dysfunction is limited. Involving well-coordinated, multicenter groups with access to experienced phase 1 investigators can significantly shorten accrual time for organ dysfunction studies. Instructions for the use CTEP organ dysfunction templates may be found at:
http://ctep.cancer.gov/protocoldevelopment/docs/renal dysfunction v3.doc

4.2　Who Is Eligible to Study Phase 1 Agents

Currently, two groups of clinical investigators are eligible to study phase 1 agents: Phase 1 U01 cooperative agreement investigators (Section 4.2.1) and qualified investigators with peer-reviewed expertise in the conduct of early clinical trials (Section 4.2.2).

Selection of phase 1 investigators is a competitive process, with preference given to those with relevant expertise, ability to correlate clinical and laboratory biologic studies, and ability to complete a high-quality study in a timely fashion. Selection of investigators is an open and competitive process and all appropriate investigators are welcome to apply.

The NIH has grant mechanisms to support junior faculty embarking upon a career in clinical research. The Career Development LOI is intended to increase the LOI success rate and to facilitate junior investigators' career development. LOIs submitted by junior investigators are prioritized favorably when compared to LOIs submitted by more experienced investigators. Since a successful outcome requires the submission of an approvable LOI, junior investigators are particularly encouraged to obtain CTEP Investigational Drug Branch staff input during LOI preparation. Information about the Career Development LOI is found on the CTEP web site, in the CTEP forms, templates and documents section.

Please refer to Appendix III for the conduct of phase 1 and 2 trials in the pediatric population.

4.2.1　Cooperative Agreement Awardees or Contractors

Phase 1 clinical trials may be supported by grants, cooperative agreements or contracts. Currently, U01 cooperative agreements support phase 1 clinical trials solicited by the NCI. Investigators are selected through competitive peer review of their cooperative agreement applications submitted in response to periodic solicitations from DCTD. These cooperative agreements are usually funded for 5 years.

When NCI seeks cooperative agreement applications, a Request for Application (RFA) is issued. Notices of the availability of RFAs are published in the "NIH Guide to Grants and Contracts" http://www.nih.gov/grants/guide/index.html and on the central web site that announces all Funding Opportunity Announcements (FOA) http://www.grants.gov/applicants/find grant opportunities.jsp.

When NCI seeks applications for a contract, a Request for Proposal (RFP) is issued. A notice of RFP availability is published in the Commerce Business Daily at http://www.gpoaccess.gov/cbdnet/ and FedBizOpps at https://www.fbo.gov/. These contract solicitations can also be found on the Office of Acquisition Management and Policy at National Institutes of Health at http://oamp.od.nih.gov/. Active contract solicitations are listed by Institute under the "Contract Opportunities" link on this web site.

4.2.2   Other Phase 1 Investigators

Qualified investigators with peer-reviewed expertise in the conduct of early clinical trials are also eligible to conduct unsolicited phase 1 trials. Investigators are usually selected because of unique expertise or research experience relevant to the agent or the availability of certain patient populations or laboratory facilities to perform special studies. These clinical studies are usually performed as investigator-initiated research using the R21, R01 or P01 grant funding mechanisms. Grants can be submitted as investigator initiated research applications or in response to a Program Announcement (PA) or Funding Opportunity Announcement (FOA). These opportunities may be found at either http://grants.nih.gov/grants/guide/index.html or http://www.grants.gov/applicants/find_grant_opportunities.jsp.

In all cases, such investigators must have demonstrated competence in conducting phase 1 studies with anticancer agents. *Ad hoc* phase 1 investigators must fulfill all CTEP requirements for trial conduct, as defined in this section, and for reporting of data as described in Section 10.

Junior faculty may submit a Career Development Letter of Intent. Please refer to Section 7.1.9.

4.3   Which Organizations Can Conduct Phase 1 Studies

Phase 1 trials are most commonly conducted by single institutions. Trials with a new single agent having an unknown adverse event profile are usually conducted most safely in a single center. In unique situations, phase 1 trials may require multiple institutions, such as with many pediatric phase 1 studies (see Appendix III), and with agents that are targeted specifically for a single disease or limited subpopulation of cancer patients. NCI Cooperative Groups are also eligible to conduct Phase 1 studies.

4.4   Who Is Eligible to Administer Phase 1 Agents

All phase 1 agents must be administered only at institutions listed on the approved protocol's cover page (except for NCI Cooperative Group trials) and must be administered under the Protocol Chair/Group Chair's supervision. Investigators must not send these agents to referring physicians. The protocol must describe in detail any part of the treatment that will be administered at a site other than the study center.

4.5   How to Obtain Information about Phase 1 Agents

4.5.1   Investigator's Brochure

This document contains all relevant information about the agent, including animal screening, preclinical toxicology, detailed pharmaceutical data, pharmacology and mechanism of action. The brochure also contains information about the clinical adverse events observed in clinical trials.   CTEP provides these routinely to investigators who are approved to conduct a clinical trial of the agent at the time the LOI is approved, and when the Investigator's Brochure is updated. When necessary, investigators with approved LOIs or protocols may obtain the Investigator's Brochure from the Pharmaceutical Management Branch (ibcoordinator@mail.nih.gov).

4.5.2   Investigational Drug Branch (IDB) Physicians

Each CTEP investigational agent is assigned to an IDB staff physician, who prepares the solicitation for initial clinical trials and coordinates its clinical development under DCTD sponsorship. Phase 1 investigators are advised to discuss a proposal with this physician before writing a formal LOI. IDB staff members welcome investigator queries prior to submission of an LOI for phase 1 trials, as well as during the subsequent development of a clinical trial protocol (see Section 7). Relevant contact information can be found in Appendix IV.

4.5.3   Other Information

Phase 1 investigators should carefully read the sections relevant to writing a protocol with phase 1 agents:

Section 7	The Letter of Intent, Concept and Drafting of a Protocol
Section 8	Protocol Review and Approval at CTEP
Section 9	Ordering Study Agents from NCI
Section 10	Responsibility for Reporting Results to CTEP
Section 12	The Investigator and Protocol Chair/Group Chair: Roles and Responsibilities
Section 15	Accountability and Storage of Investigational Agents
Section 16	Monitoring and Quality Assurance

The following section explains the policies for phase 2 clinical investigations with CTEP study agents. We outline the scientific objectives for CTEP-supported clinical trials and describe which physicians are eligible to study and administer study agents. These two aspects of agent use-study and administration-are formally separate issues. In general, eligibility to study CTEP study agents is restricted to institutions and investigators approved by peer review; these individuals may include grantees, contractors, Cooperative Group members, physicians affiliated with approved cancer centers, and recipients of investigator-initiated clinical research project grants (RO1, PO1). Except in certain explicitly defined circumstances outlined in the following sections, the administration of study agents is restricted to these investigators.

For investigational agents involving a pharmaceutical collaborator and a DCTD-sponsored IND, the Investigational Drug Steering Committee (IDSC) is consulted to develop a strategic clinical development plan. The IDSC membership includes Principal Investigators from NCI Phase 1 U01 grants and Phase 2 N01 contracts, as well as representatives from Cooperative Groups. Its members include experts in biostatistics, imaging, radiation oncology, and clinical and preclinical pharmacology. The IDSC subgroups assist CTEP with early phase trial prioritization, and address critical scientific issues for the trials
(http://ctep.cancer.gov/SpotlightOn/Investigational Drug Steering Committee.htm).

5.1   New Agent Development Considerations

5.1.1   Planning and Coordination of Phase 2 Trials by CTEP

As a sponsor, DCTD must devise and implement plans for phase 2 trials of novel therapeutics. An adequate phase 2 plan, while conceptually straight-forward, is often difficult to execute. A reasonable plan presupposes answers to the following questions:

- What doses and schedules emerging from phase 1 ought to be carried forward into phase 2?
- What diseases should be targeted for testing?
- How does the new agent fit into CTEP's priority list for various targeted disease studies?
- How does the new agent fit into the priorities of the clinical investigators who form the core of the NCI-supported clinical trials network?
- How can the CTEP assure that each agent is adequately tested in each disease that is studied? How many studies should be mounted in each disease category? What kinds of patients are suitable for study entry? What are suitable stopping rules for phase 2 trials?
- How should we perform phase 2 studies if there are limited supplies of the new agent?
- What important laboratory correlates can be made within the context of a clinical trial?
- How can the proposed study be completed within a suggested timeline (i.e. multicenter vs. single center)?

CTEP staff collaborates with each agent's industrial sponsor and the Investigational Drug Steering Committee (IDSC) during late phase 1 to plan phase 2 development; CTEP announces the plan via solicitation of LOIs.

5.1.2   Single Agent Phase 2 Studies

A phase 2 study:
- determines whether an agent has antitumor activity and
- estimates the response rate in a defined patient population.

Well-designed phase 2 trials limit enrollment to just the number of patients needed to ensure detection of a medically significant level of activity.

Phase 2 studies are disease-oriented. Various tumor types are tested in phase 2 as distinct clinical entities, as each has differing prognostic factors, eligibility requirements, and patterns of responsiveness to a particular agent. As there may be many unknown or uncontrollable factors contributing to variability in outcome, CTEP attempts to sponsor two phase 2 trials in each tumor type.

The goal of these initial phase 2 trials is to determine whether the new agent has activity against particular cancers. These trials, therefore, serve as a screen for further study. For this reason, investigators must make every effort to avoid false results. Although false-positive results are certainly undesirable, false-negative phase 2 results are also damaging, as they may delay significantly or prevent discovery of a potentially useful antitumor agent.

CTEP has based its guidelines concerning eligibility requirements on patient characteristics that appear to have a particular impact on likelihood of response. Specifically, for initial phase 2 studies, we currently seek trials that restrict patient eligibility to the minimum extent of prior therapy consistent with current medical practice. If an agent's MTD is well characterized, protocols for its initial phase 2 trials should restrict patient entry in the following ways:
- For diseases that currently lack effective systemic therapy (e.g. liver and pancreas), trials should be limited to patients with no prior chemotherapy.
- For diseases in which systemic therapy may cause objective tumor regression but has little or no impact on survival, entry of patients with no prior therapy will also be sought, whenever possible (e.g. carcinomas of the head and neck, cervix, esophagus, prostate, bladder, large bowel, kidney, stomach, non-small cell lung, and melanoma).
- For diseases that are potentially curable with systemic treatment (e.g. acute leukemias, diffuse non-Hodgkin's lymphomas, Hodgkin's disease, testicular cancer, limited small cell lung cancer, and ovarian cancer), patients having the minimum extent of prior treatment consistent with current ethical standards of care are selected.

This policy will have the following desirable consequences:
- Patients initially entered into phase 2 trials will have the best chance of benefit from treatment and should be able to tolerate any adverse effects of therapy better than patients with poorer performance status, agent-resistant disease, and possible compromised major organ function from prior chemotherapy.

- Fewer patients are exposed to inactive agents.
- The chance of missing potentially active agents will be minimized.

Clearly, the population of patients defined in this way is highly selected, and the results of these initial trials will not necessarily represent the agent's activity in the general population of patients with the disease in question. Once a new agent shows significant activity in this initial, relatively favorable, subset of patients, eligibility criteria in subsequent studies will permit entry of patients with less favorable prognostic characteristics, so that such patients may have an opportunity to benefit from an active agent. In this second stage of the new agent's phase 2 evaluation, a more accurate assessment of its activity in the general population of patients with cancer may be obtained.

If an agent shows promising anti-tumor affects, CTEP may perform a "Special Response Review." The audit may be conducted at CTEP or on-site. Auditors are selected to participate according to their expertise. Pharmaceutical representatives and IDB staff may also be present. Selecting only responding patients, the investigator(s) presents a written and oral summary and review of each claimed responding patient. The coordinating site is responsible for providing all materials, such as MRIs, scans and all tumor measurements according to the response criteria and evaluation schedule stated in the protocol. CTMB prepares a full report including patient summaries, tumor measurements and endpoint results. The CTMB-prepared final report is approved by the lead expert auditor and the CTEP agent or disease coordinator. The PI may include a statement in subsequent manuscripts that CTEP has verified the responses. This statement may not be used if all claimed responding patient's results were not verified.

5.1.3   Combining Agents

A rational approach to development of a new cancer agent in combination with another agent(s) may include (1) the demonstration of evidence of single agent activity of each agent in a disease, and/or (2) convincing laboratory evidence that is relevant to clinical circumstances.

In the past, oncologists developed many agent combinations intuitively; they combined two, three, or more putatively active agents in uncontrolled studies of antitumor effect and adverse effects. Some very real therapeutic advances were achieved by this process. However, the lack of a systematic, stepwise approach and the frequent absence of proper control groups often left the oncology community in the uncertain position of not knowing whether results with a particular regimen represented progress or not. A new agent's overall impact on efficacy and adverse events may remain unclear without a systematic approach. Ultimately, when available data do not elucidate each new agent's specific contribution, the process of NDA or BLA approval is impeded.

Intelligently designed and flexible new agent development programs must provide room for both approaches. Well-conceived small pilot trials testing new hypotheses will always have an important place in cancer's developmental therapy. We shall, however, continue to pay close attention to the rationale behind all proposed combinations. We will also continue to ask whether certain proposals for therapeutic research might not be better approached by a phase 3 rather than phase 2 design.

In the past, activity was the most common basis for a single agent's inclusion in a combination. CTEP now considers other rationales as well. For example, we consider radiosensitizers or substantial laboratory evidence of synergy between two cancer agents. This is particularly compelling if it is consistent with a putative mechanism of action. Alternatively, an agent inert against cancer might be added because of evidence that it alters a second (anticancer) agent's pharmacodynamics or pharmacokinetics. Agents can also be tested as both chemotherapeutics and radiosensitizers. In such cases, CTEP will carefully assess the rationale and evidence offered in support of a proposal.

When investigators submit combination studies to CTEP for review, therefore, it is particularly important to state the proposal's goals, background, and rationale clearly.
- If experimental results in the laboratory are the basis for the study, they should be relevant to the clinical circumstance and cited in adequate detail.
- If preliminary clinical results are the motivation, they should be similarly cited; unpublished results should be provided as part of the background or in an attachment to the protocol document.
- If the trial proposes a feasibility pilot, the protocol should state clearly what kinds of results the investigators would regard as medically significant and where they would propose to go next if a significant result were obtained.

A detailed plan of a follow-up study or detailed speculations about likely outcomes is not necessary at this stage of review. Rather, we are seeking an understanding of how the pilot proposal will fit into a strategy of development of the new therapeutic idea.

5.1.4   Randomized Phase 2 Studies

A randomized controlled trial is the study design that can provide the most compelling evidence that the study treatment causes the expected effect on human health. This study design has become a common practice to conduct "active comparator" studies (also known as "active control" trials). In other words, when a treatment exists that is clearly better than not treating the subject (*i.e.,* giving them a placebo), the alternate treatment would be a standard-of-care therapy. The study would compare the "test" treatment to standard-of-care therapy. Comparison to a control arm is most useful when there is little prior information on expected efficacy rates and can also be useful for end points that can be heavily influenced by patient selection, such as Time to Progression (TTP) and Progression Free Survival (PFS). However, except in very rare cases, a larger phase 3 study will be required to definitively establish clinical benefit. Although it will not be definitive, the phase 2 randomized control design will often aid decisions on whether to pursue further study of the treatment and guide the design of additional trials.

Appropriate data and safety monitoring plans are required to be in place for all phases of therapeutic clinical trials supported by CTEP that are consistent with NIH and NCI policy as explained on the CTEP website under "Data and Safety Monitoring (DSM) Plans" under the CDE / Data Policies / CDUS section at: http://ctep.cancer.gov/protocolDevelopment/default.htm#cde_data_pol_cdus. This is to insure the safety of participants, the validity of data, and the appropriate termination of studies for which significant benefits or risks have been uncovered or when it appears that the trial cannot be concluded successfully. These plans must be in place before grants or contracts for clinical trials organizations supported by NCI/CTEP are awarded and/or detailed in the protocol document. Particular studies, including randomized phase 2 trials, may require more specific monitoring and special requirements may be provided in the protocol document, including how interim results will be monitored by any Data Monitoring Committee overseeing the trial. For therapeutic randomized phase 2 trials supported by NCI/CTEP, CTEP may require that these studies be monitored by a specific Data and Safety Monitoring Board (DSMB). Cooperative Groups should follow the policy used for monitoring all Cooperative Group phase 3 trials. This policy is described under "NCI Cooperative Group Data Monitoring Committee Policy" under the CDE / Data Policies / CDUS section at: http://ctep.cancer.gov/protocolDevelopment/default.htm#cde_data_pol_cdus.

More information on NCI policies regarding data and safety monitoring of clinical trials is available at http://grants.nih.gov/grants/policy/hs/data_safety.htm.

5.2   Protocol Considerations

5.2.1   Single Disease Studies

Each tumor type should be considered for phase 2 study separately. In general, this means that there should be separate protocols for each tumor type. If a compelling reason supports including several under one protocol, such as uncommon tumors, then investigators should include separate criteria for each tumor type regarding: (a) eligibility requirements, including extent of prior treatment, (b) acceptable sites for measurable disease, (c) response assessment, and (d) accrual objectives.

5.2.2   Eligibility Requirements

- Tumor Types: For each proposed tumor type there should be separate criteria for eligibility.
- Prior Therapy: Because it is clear that the extent of prior cytotoxic chemotherapy is an important determinant of response probability, CTEP seeks initial trials that restrict patient eligibility to minimal prior therapy that is still consistent with good medical practice. (see Section 5.1.2)
- Measurability of Disease: To define quantitatively the antitumor activity of an agent, patients in phase 2 trial must have measurable disease parameters.

In certain diseases, common sites of involvement are either not bidimensionally measurable or assessment techniques do not permit quantifiable measurement. Under these circumstances, investigators may evaluate tumor response without quantification; in such cases, having more than one observer assess responses is particularly desirable. Examples include bone metastases, lymphangitic pulmonary disease, and many parenchymal brain lesions.

Performance Status:
Under most circumstances, entry to initial phase 2 studies should be confined to patients who are largely ambulatory (ECOG $\leq$ 2). Patients should be expected to survive a sufficient period of time for adequate observations to be made.

Organ Function:

Evidence that major organ function is normal is required since impairment would compromise the safe use of the investigational agent.

Gender:

Where appropriate, the study should enroll both women and men. NIH policy requires that women and members of minority groups must be included in all NIH-supported biomedical and behavioral research projects involving human subjects, unless a clear and compelling rationale and justification establishes inclusion is inappropriate with respect to the health of the subjects or the purpose of the research. Please include a separate section regarding the "Inclusion of Women and Minorities" that describes the inclusion of women and members of minority groups appropriate to the study's scientific objectives. (http://grants.nih.gov/grants/funding/women_min/women_min.htm) Within the protocol, the investigators must describe the proposed study population's composition in terms of gender and racial/ethnic group, and provide a rationale for selecting such subjects. The investigator must include a table in the protocol text.

5.2.3   Accrual and Statistical Considerations

Investigators should specify each study's accrual goals in advance, and state the proposed maximum number of patients explicitly. They must provide justification for the target sample size, in terms of precision of estimation or levels of type I and type II error. CTEP recommends multistage designs for distinguishing an unacceptable level of response from a promising level {e.g., Fleming, *Biometrics* 38:143 (1982); Simon, *Controlled Clinical Trials* 10:1, (1989)}. Investigators should anticipate the accrual rate of eligible patients *realistically*, and describe mechanisms that are in place for early stopping of negative trials.

For cases where randomized phase 2 studies are preferable, Rubinstein LV et al. (Journal of Clinical Oncology 2005 23: 7199) describes an exploratory study design ("randomized phase 2 screening designs") which may be useful. This design applies, in particular to situations where there is interest in adding an experimental agent to a standard therapy for a particular cancer. The design would facilitate conducting a trial comparing regimen A (standard therapy) to regimen A plus the new agent.

5.3   Which Organizations can Conduct Phase 2 Studies?

5.3.1   Cooperative Groups

All registered physicians who are members of a Cooperative Group, including those at full member institutions, in Community Clinical Oncology Programs (CCOP) (http://prevention.cancer.gov/programs-resources/programs/ccop), or at affiliate institutions or special members may participate as investigators on CTEP's phase 2 and phase 3 trials. However, a Cooperative Group may have policies that restrict investigators' participation.

5.3.2   Cancer Centers

Physicians with current/active registration with CTEP at institutions designated as comprehensive or clinical Cancer Centers by the NCI may participate on CTEP's phase 2 and phase 3 trials. Such physicians may be:
- Staff physicians within the Center;
- Physician members of CCOPs for which that center is the research base; and
- Physicians affiliated with Cancer Centers (see Section 13 for further details on the affiliate policies of CTEP).

5.3.3   Affiliates

Physicians affiliated with a clinical site may participate as investigators on CTEP's phase 2 and 3 clinical trials provided that:
- The affiliation is formalized and its terms are in writing and based on the CTEP policies on affiliates. (see Section 13);
- Each investigator is registered and active with CTEP by annual submission of a signed FDA Form 1572, Supplemental Form for Investigator Registration, and Financial Disclosure Form, and CV (see Section 12 and http://ctep.cancer.gov/forms/index.html).

5.3.4   New Agent Development Contractors and Cooperative Agreement and Grant Awardees

This category includes those with a Phase 1 U01 grant, a Phase 2 N01 contract, U10 cooperative agreement award and other NCI-funded consortia including: (a) the Adult Brain Tumor Consortium, (b) the AIDS Malignancies Clinical Trials Consortium, (c) the Pediatric Phase 1 Clinical Trials Consortiums and (d) the Pediatric Brain Tumor Consortium. This category also includes investigator-initiated grants to study new agents (e.g. R01, R03, R21 and P01).

5.3.5   Multicenter Phase 2 Trials

CTEP expects that many phase 2 trials will be performed only at the proposing clinical site. If a Protocol Chair wishes to collaborate with other institutions not formally affiliated with his or her clinical site, the protocol should include a description of the procedures by which the collaborating institutions will manage the conduct of the protocol. The investigator should list each institution and the name of responsible investigator at each on the protocol face sheet. The protocol should specifically address issues described in Section 7.2.17. If a Cooperative Group is leading a phase 2 trial, the Group should specify if it is an institution-limited study or if it is open to its entire membership (assuming the individual sites meet any study-specific requirements).

5.4   Who is Eligible to Administer Phase 2 Agents

For a particular clinical protocol, physicians who may prescribe and administer DCTD investigational agents are:
- those registered and active with CTEP (see Section 14.1);
- members of any research base or formally designated affiliate listed on the protocol's face sheet; and
- others who are individually named on the protocol's face sheet.

5.5   Restriction on Participation in Phase 2 Studies

CTEP may restrict the testing of any investigational agent to a limited number of locations. Although most new agents proceed to a phase 2 program open to all eligible investigators, some are restricted to single centers, specific centers or specific investigators. Such restrictions may remain until a safe, reliable phase 2 dose has been defined and CTEP and the investigator community are confident that the agent is ready for general testing among all investigators. Inadequate agent supply may also prompt restrictions.

5.6   How to Obtain Information about Phase 2 Agents

5.6.1   Investigator's Brochure

This document contains all relevant information about the agent, including animal screening, preclinical toxicology, detailed pharmaceutical data, pharmacology and mechanism of action. The brochure also contains information about clinical adverse events observed in clinical trials. CTEP provides these routinely to investigators who are approved to conduct a clinical trial of the agent at the time the LOI is approved, and when the Investigator's Brochure is updated. When necessary, investigators with approved LOIs or protocols may obtain the Investigator's Brochure from the Pharmaceutical Management Branch (ibcoordinator@mail.nih.gov).

5.6.2   IDB Physicians

CTEP assigns each investigational agent to an IDB staff physician, who is responsible for coordinating the agent's clinical development. Investigators contemplating submitting a clinical trial proposal are encouraged to contact the IDB physician(s) responsible for the investigational agent(s).

5.6.3   Clinical Research Pharmacists

The Pharmaceutical Management Branch has a staff of clinical research pharmacists who interact closely with IDB and Clinical Investigations Branch (CIB) staff physicians. Clinical research pharmacists are available to provide pharmaceutical and agent information data on DCTD investigational agents. (See Appendix IV for contact information).

The following section explains the policies for phase 3 clinical investigations with CTEP study agents. We outline the scientific objectives for CTEP-supported clinical trials and describe which physicians are eligible to study and administer study agents. Except in certain explicitly defined circumstances outlined in the following sections, the administration of study agents is restricted to these investigators.

6.1   Scientific Policies of CTEP

If investigators observe significant activity with an experimental therapy in any disease during phase 2, further clinical trials usually compare the experimental therapy's efficacy with that of a standard or control therapy. If reasonable standard treatment can be defined for the disease in question, we generally wish to know whether the new agent or therapy constitutes a significant contribution in terms of patient benefit. A variety of trial designs may be suitable, according to the standard of care treatment in the particular disease. Those that are most satisfactory are controlled trials that compare the new agent to a standard single agent or a standard regimen plus the experimental agent to the standard regimen alone. Regardless of the design selected, however, an appropriate control group must exist and relevant endpoints must be used to measure relative effects. Of greatest medical importance are relative survival and quality of life. Other measures, such as complete remission rate or disease-free survival, may also be of interest.

These studies, which attempt to isolate a new agent's role in the treatment of a specific cancer, are of obvious importance to pharmaceutical collaborators because the results are pivotal in applications to register the agent for commercial distribution. They are of equal importance to the oncology community, because such approval makes the agent generally available for patient care. These trials' results may also be of great medical importance. If the control group is properly selected such trials may yield valuable information for the care of cancer patients.

Every protocol must contain a section that discusses the study design and the plan for data analysis. The protocol should include a statement regarding the study's major objectives as hypotheses to be tested, and a clearly specified target sample size. The sample size goal should be justified in terms of precision of estimation or based on levels of type I and type II error. In a phase 3 study, it is insufficient simply to give the number of patients to be accrued on each arm. The protocol should specify the test to be used to compare the treatment groups, and the probabilities of drawing incorrect conclusions when performing this test with the proposed sample size. The magnitude of improvement in outcome that can be reliably detected using the planned sample size should also be specified. The accrual rate of eligible patients per year that can be realistically anticipated should be stated and documented. The protocol should describe specific statistical plans for interim analysis of accumulating data. Therapeutic randomized phase 3 Cooperative Group trials supported by NCI are required to have a Data Monitoring Committee (Data and Safety Monitoring Board) with organization, responsibilities and operations consistent with Cooperative Group Data and Safety Monitoring Board policy. CTEP policies are available on the CTEP website under "NCI Cooperative Group Data Monitoring Committee Policy" under the CDE / Data Monitoring Policies / CDUS section at: http://ctep.cancer.gov/protocolDevelopment/default.htm#cde_data_pol_cdus. NCI policies are available at http://www.cancer.gov/clinicaltrials/patientsafety/dsm-guidelines/page2. The plan for the Data Monitoring Committee to monitor interim results should be indicated in the protocol.

If evaluation of treatment effect will require use of nonrandomized controls, a thorough description of the control group to be employed should be part of the protocol. This description should include a detailed discussion of comparability issues and analytic techniques.

6.2   Who Is Eligible to Conduct Phase 3 Trials

CTEP-supported phase 3 therapeutic trials are conducted via the NCI-funded Clinical Trials Cooperative Group Program. Cooperative Groups submit phase 3 study proposals ("concepts") to the NCI disease-specific Steering Committee that oversees the particular disease area and/or CTEP which evaluates and approves the phase 3 trial concept. Because sample sizes required for such studies are usually quite large, a multicenter approach is usually the only feasible way to conduct such a trial. NCI-funded investigators who are not associated with a Cooperative Group who have a potential phase 3 study idea are encouraged to discuss the idea with their representatives on a disease-specific Steering Committee (SC) and/or Task Forces associated with the SC or CTEP (if no disease specific SC exists in that disease). It is anticipated that promising phase 3 study ideas would move forward via collaboration with a Cooperative Group.

Proposals for phase 3 studies should document very specific accrual potential. The protocol should include a description of procedures by which the collaborating institutions or trial organizations will manage the conduct of the protocol.

On occasion, a Cooperative Group may wish to join a phase 3 trial being conducted by an outside, established, independent clinical trials organization (e.g., European Organisation for Research and Treatment of Cancer (EORTC)). These studies must also undergo full scientific review by the appropriate disease-specific SC and/or CTEP as well as approval by CTEP regarding scientific resource, regulatory, operational, and logistical issues related to the collaboration for consistency with the requirements of the Cooperative Agreement for Group trials before such a collaboration is approved.

6.3   Eligibility Requirements

Phase 3 clinical trials must include a review of the available evidence to show whether or not clinically important gender or race/ethnicity differences in the response to the intervention are expected. The trial's design must reflect the current state of knowledge about expected differences. Phase 3 clinical trials are, in addition, required to provide valid analysis to measure differences of clinical or public health importance in intervention effects based on gender or racial/ethnic subgroups where evidence supports differences.

Investigators should consider the following circumstances when planning a phase 3 clinical trial and include the appropriate anticipated accrual table information in the protocol document:

- Prior data strongly indicate that the intervention will show significant clinical or public health differences among gender, racial, and/or ethnic subgroups. In this case, the proposed phase 3 trial's primary question(s) to be addressed and design must specifically accommodate these differences. For example, if men and women are thought to respond differently to an intervention, the phase 3 trials must be designed to answer two separate primary questions, one for men and the other for women, with adequate sample size for both.

- Prior data strongly support no significant clinical or public health differences among subgroups from the intervention. In this case, gender, race, and/or ethnicity will not be required as subject selection criteria. However, the inclusion of gender, racial, and/or ethnic subgroups is still strongly encouraged.

- Prior data neither strongly support nor negate the existence of significant clinical or public health differences among groups. In such cases, the phase 3 trial must include sufficient and appropriate gender, racial, and/or ethnic subgroups, so that valid analysis of the intervention effects on subgroups can be performed. However, the trial will not be required to provide high statistical power for each subgroup.

Cooperative Group Phase 3 studies:
Effective October 1, 1995, all phase 3 protocols must include accrual targets for males, females, and minorities, reflecting expected accrual over the life of the study. The NCI suggests basing accrual targets on data from similar trials completed by the Cooperative Group during the previous 5 years. Accrual targets should resemble the gender, racial, and ethnic composition of the U.S. population as closely as possible. A worksheet, including a description of the currently recognized HHS racial and ethnic categories, is attached for your reference. (http://ctep.cancer.gov/branches/cgcb/gender_minority.htm)

CTEP will return protocols that fail to address the above gender and minority issues without Protocol Reviewed Committee (PRC) review.

Planned Gender and Minority Inclusion:

Accrual Targets

Ethnic Category	Females		Males		Total
	Females		**Males**		**Total**
Hispanic or Latino		+		=	
Not Hispanic or Latino		+		=	
Ethnic Category: Total of all subjects	(A1)	+	(B1)	=	(C1)
Racial Category					
American Indian or Alaskan Native		+		=	
Asian		+		=	
Black or African American		+		=	
Native Hawaiian or other Pacific Islander		+		=	
White		+		=	
Racial Category: Total of all subjects	(A2)	+	(B2)	=	(C2)
	(A1 = A2)		(B1 = B2)		(C1 = C2)

Accrual Rate: _____ patients/month   **Total Expected Accrual:**_____ Min _____ Max

Projected Start Date of Study:_____

HHS Racial and Ethnic Categories

I. **American Indian or Alaskan Native:** A person having origins in any of the original peoples of North America, and who maintains cultural identification through tribal affiliation or community recognition.

II. **Asian or Pacific Islander:** A person having origins in any of the original peoples of the Far East, Southeast Asia, the Indian subcontinent, or the Pacific Islands. This area includes China, India, Japan, Korea, the Philippine Islands, and Samoa.

III. **Black, not of Hispanic Origin:** A person having origins in any of the black racial groups of Africa.

IV. **Hispanic:** A person of Mexican, Puerto Rican, Cuban, Central or South American, or other Spanish culture or origin, regardless of race.

V. **White, not of Hispanic Origin:** A person having origins in any of the original peoples of Europe, North Africa, or the Middle East.

Special Populations:
Individuals from special populations (minorities, cancer survivors, HIV+ individuals, pregnant and breast-feeding women) can NOT be arbitrarily excluded from participation on a study. All exclusions must be justified based on establishment that inclusion is inappropriate with respect to the health of the research subjects or the purpose of the research.

6   Phase 3 Trials

6.4   Coordination of Planning with the Clinical Investigations Branch (CIB)

The Clinical Investigations Branch (CIB) of CTEP is responsible for scientific oversight and coordination of large, multicenter clinical trials exploring innovative disease therapeutics and biomarkers. CIB partners with public and private entities to expand clinical trial participation to all populations. Large clinical trials involve years of effort and a substantial expenditure of resources. Accordingly, a certain amount of coordination is necessary for the optimal planning of specific studies of this type.

CIB staff are well positioned to advise investigators about the existence of other proposed or ongoing studies that are closely related or even identical to studies being planned. They can also advise about overall disease-specific goals. Although investigators proposing phase 1 and phase 2 studies using CTEP resources are required to submit a Letter of Intent (LOI), for phase 3 studies, CTEP requires investigators to submit a written Concept using the Concept Submission Form (http://ctep.cancer.gov/forms/docs/concept_submission.doc). The NCI disease-specific Steering Committee overseeing that disease area must evaluate Cooperative Group Concepts for large phase 2 studies (100 or more patients) and all phase 3 studies. In addition, the disease-specific Steering Committees review all phase 2 concepts from non-Group investigators over 100 patients that request CTEP IND agents, with the exception of LOIs submitted in response to a CTEP mass solicitation. The LOIs submitted in response to a mass solicitation are reviewed by the CTEP Protocol Review Committee (PRC) and not by the Steering Committees. Also, in the event that there is a concept for a large phase 2 study in a disease for which there is not a Steering Committee, then CTEP PRC would conduct the review.

CTEP may also refer Cooperative Group phase 2 trials with total sample sizes of 90 to 99 patients that are not being submitted in response to a specific CTEP solicitation to the appropriate NCI disease-specific Steering Committee for evaluation, so the Cooperative Group for these trial proposals should also submit them to CTEP using the Concept Submission Form.

In general, the Concept should describe the proposed study, including the hypothesis to be investigated, its rationale, and relevant design considerations. The appropriate NCI disease-specific Steering Committee and/or CTEP then formally evaluates the study proposal and provides a written Concept Evaluation on the study proposal's scientific merits and feasibility.

6.5   Cancer Trials Support Unit (CTSU)

In response to the Armitage Report, a 1997 report from the NCI's Clinical Trials Program Review Group, the Cancer Trials Support Unit (CTSU) was established to:
(1) facilitate physician and patient access to NCI-sponsored clinical trials through an efficient enrollment procedure that helps cross-Group accrual and permits non-Group members to enroll patients on NCI-sponsored trials;
(2) streamline data entry and collection for clinical trials using standard case report forms and reporting; and
(3) reduce regulatory and administrative burdens on clinical trials by unifying and standardizing Group membership rosters and institutional review board (IRB) approvals.

6   Phase 3 Trials

In general, the CTSU clinical trials menu includes all phase 3 clinical therapeutic trials conducted by the Cooperative Groups. This allows the members of any Cooperative Group to participate in any phase 3 trial (and selected, large, phase 2 trials) led by a Cooperative Group that is available on the CTSU clinical trials menu. In addition, the CTSU has other initiatives; including a pilot program supporting trials for other NCI-sponsored clinical trials networks and a centralized patient enrollment system (OPEN) for all Cooperative Group study enrollments.

Additional information on the CTSU can be found on its web site at https://www.ctsu.org/.

6.6   Non-Clinical Studies Using Trial Specimens or Clinical Data

Investigators may also submit non-clinical study proposals to CTEP for review that include requests to use specimens with or without associated clinical data from therapeutic clinical trials which involved a CTEP-provided Agent (e.g., request to use banked biospecimens from a Cooperative Group trial for a correlative science study). These studies must be reviewed and approved by CTEP or a CTEP recognized disease-specific Correlative Science Review Committee. Release of specimens and any associated clinical data for these non-clinical trials must also be in compliance with the CTEP IP Option available on the CTEP website at: http://ctep.cancer.gov/industryCollaborations2/default.htm#guidelines_for_collaborations .

Company collaborators for the therapeutic clinical trial for which specimens and associated clinical data are being requested will also be notified of these requests and may submit comments on the non-clinical study proposal.

Planning and Execution of a Clinical Trial

The following five sections describe the investigator's responsibilities for implementation of a clinical trial, from drafting the protocol to study completion. They guide the Protocol/Group Chair and participating investigators and outline NCI policies on their responsibilities in clinical trial execution.

7.1   The Letter of Intent (LOI) or Concept

7.1.1   Definition

The Letter of Intent (LOI) is an investigator's declaration of interest in conducting a phase 1 or 2 trial with a specific investigational agent(s). CTEP's approval of the LOI reserves that "slot" for the investigator's protocol if it is submitted within a defined time frame. Approval also signifies agreement that the investigator shall submit a protocol based on the terms stated in the LOI. The Concept serves the same purpose for Cooperative Group phase 3 studies and large, phase 2 trials that are not submitted in response to a CTEP mass solicitation, as described in Section 6.4. Investigators are encouraged to contact the CTEP IDB staff prior to submitting an LOI for a CTEP IND agent. They may learn, for example, whether their proposal would be duplicative or whether there are sufficient preliminary data to move ahead with a clinical trial.

7.1.2   Purpose

CTEP has devised the LOI and Concept system to maximize the efficiency and fairness by which it allocates experimental agents to investigators for study. Proper use of the system ensures a steady flow of new agents into the clinical trials system. It enables CTEP to plan the development of several agents simultaneously. For investigators, the LOI/Concept system also saves time and effort, because its use will spare them from writing a protocol unlikely to be approved. Protocols submitted subsequent to favorable review of an LOI/Concept are much more likely to be approved without request for major modification, because many of the crucial features of the trial must be specified in the LOI/Concept itself. The LOI/Concept system also is used for submission of combination pilot studies. In these cases, reviews typically focus on the rationale for agent combinations, the proposed sample size, and the adverse events of each agent when given alone. The system also provides the investigator with an opportunity to explore the proposal with CTEP staff at an early stage.

7.1.3   Ground Rules for the LOI and Concept System

Investigators must submit LOIs for phase 1 or 2 trials that include a CTEP investigational agent according to the following schedule:
Agents Beginning Phase 1 - In advance of the IND submission, CTEP will announce the availability of an agent, issue a request for proposals (solicitation) for phase 1 trials and provide a deadline for the submission of LOIs.
Agents Beginning Phase 2 - In late phase 1, CTEP will issue a request for proposals for initial phase 2 trials including a deadline for submission of LOIs.
All Other Phase 2 Trials - After this deadline has passed, investigators may submit LOIs (or Cooperative Group Concepts for large phase 2 trials) at any time.

Cooperative Groups that propose phase 2 trials, as defined in <u>Section 6.4</u>, should submit their study proposals as Concepts instead of LOIs if they are not being submitted in response to a specific CTEP solicitation. These large phase 2 trials will be evaluated by the appropriate NCI disease-specific Steering Committee (SC) or CTEP, if an NCI disease-specific SC is not operational in a specific disease area.

Phase 3 trial proposals must always be submitted as Concepts.

Each phase 1 or 2 protocol must be preceded by an approved LOI (or Concept). If a phase 1 or 2 protocol is submitted without a prior LOI (or Concept) having been submitted and approved, CTEP will notify the proposing investigator(s) that CTEP will not review the submission. Please refer to http://ctep.cancer.gov/protocolDevelopment/letter_of_intent.htm for additional information on LOIs/Concepts. In a similar manner, phase 3 protocols must also be preceded by an approved Concept.

7.1.4   Submission of LOIs/Concepts

In order to review the LOI properly, CTEP must have the following information:
- Principal Investigator (PI);
- Lead Group/Institution;
- Other Participating Groups/Institutions;
- Requested CTEP study agent(s);
- Other study agents to be used on the trial and their source;
- Tumor type;
- Patient characteristics, including extent of prior therapy, performance status, and abnormal organ function permitted (if any);
- Phase of study;
- Treatment plan–Agents, doses and schedule of administration;
- Rationale/hypothesis;
- Proposed correlative studies;
- Endpoints/statistical considerations;
- Proposed samples size;
- Estimated monthly accrual;
- Accrual documented by prior (similar) trials; and
- List of competitive studies
- PI Opt In/Out of the Investigational Drug Steering Committee (IDSC) process for phase 1 or phase 2 studies

Investigators should provide this information on the LOI Submission Form found on the CTEP web site http://ctep.cancer.gov/protocolDevelopment/letter_of_intent.htm#instructions and submit the completed form electronically to the CTEP Protocol and Information Office (PIO), at pio@ctep.nci.nih.gov. Cooperative Group Investigators should provide this information on the Concept Submission Form found on the CTEP web site http://ctep.cancer.gov/protocolDevelopment/default.htm#lois_concepts for large phase 2 studies of 100 patients or more (or phase 3 studies) and submit it electronically to the CTEP PIO at pio@ctep.nci.nih.gov. Task Forces associated with an NCI disease-specific Steering Committee should have already discussed Phase 3 study proposals prior to their submission to PIO, CTEP.

The Protocol Chair and the Group Chair or his/her explicit designee (e.g. Executive Officer) must cosign Cooperative Group LOIs/Concepts. The U01/N01 Principal Investigator must cosign any LOIs submitted under their contract.

We encourage Protocol/Group Chairs to submit, where appropriate, a letter with the LOI/Concept that explains the rationale, where not obvious, or any unique features of the study in greater depth. Such additional explanation is not usually necessary for single agent phase 2 trials but may assist in the review of more complex proposals involving experimental agents.

7.1.5   Review of LOIs/Concepts

On receipt of a LOI/Concept, CTEP sends an acknowledgment to the investigator. The CTEP Protocol Review Committee (PRC) reviews LOIs. CTEP-approved LOIs are sent to the pharmaceutical collaborator for review and agent commitment. Pharmaceutical collaborators are asked to respond within a few weeks of receipt. An approval or disapproval letter is issued to the Principal Investigator usually within approximately 30 days of submission. NCI disease-specific Steering Committees evaluate Cooperative Group large phase 2 study Concepts as well as phase 3 trial Concepts (or CTEP if no NCI disease-specific Steering Committee exists in a disease) and issue a letter of approval or disapproval to the Cooperative Group usually within 45 days of submission. These may also require pharmaceutical collaborator review and agent commitment. CTEP (with the NCI disease-specific Steering Committee, as appropriate) may also issue a "hold" letter for LOIs or a "pending" letter for Concepts if particular issues related to the proposal need to be addressed before a final decision can be made.

7.1.6   CTEP LOI/Concept Review Criteria

At the time of LOI/Concept review, the PRC (or NCI disease-specific Steering Committee) requires information on other investigators' studies in that agent/disease combination, as well as a listing of other studies the submitting investigator is conducting in the proposed disease. The committee considers the following in its deliberations:
- The rationale for the study (especially for combinations of agents);
- Study design, including dose, schedule, and comparison groups, if relevant;
- The study population's characteristics, particularly the extent of prior chemotherapy and performance status;
- The feasibility of the projected accrual, including an assessment of the investigator's past performance in that tumor type;
- Competing studies of the investigator in that disease;
- All other protocols and LOIs for that agent/disease combination from other sources; and
- Any features unique to the proposal

7.1.7   Targets and Deadlines for LOI/Concept Submission to Protocol Activation

Following approval of an LOI or Concept, the LOI Principal Investigator or Cooperative Group Chair has approximately 60 days to submit a protocol that conforms to the plan agreed to at the LOI stage of development. Timelines for LOI/Concept review, protocol development, and trial activation were agreed to as part of the NCI's Operational

Efficiency Working Group (OEWG) for implementation as of January 1, 2011. Target timelines, as well as "absolute" deadlines for LOI/Concept review, protocol development, and trial activation were established and are available at http://ctep.cancer.gov/SpotlightOn/OEWG.htm.

7.1.8   Information about the Status of an LOI/Concept

Further information about the status of a particular LOI may be obtained by contacting the CTEP LOI/Concept Coordinator (pio@ctep.nci.nih.gov Attn: LOI/Concept Coordinator).

7.1.9   CTEP Career Development LOI

The Career Development LOI is intended to increase the LOI success rate and to foster junior investigators' career. The process includes LOI prioritization and documentation of mentorship and institutional support.

Eligibility:

- The PI should have a major interest in and intend to develop a career in clinical research.
- He/she should be within 7 years of completion of fellowship training and a faculty member (fellows may not serve as PIs on studies) at an institution with a successful track record in conducting cancer clinical trials (Note: PIs of NCI Cooperative Group proposals are also eligible).

To submit the LOI as a Career Development LOI, the PI curriculum vitae, institutional letter of commitment, and mentor letter of commitment should be attached to the LOI. Additional information and instructions regarding the submission of a Career Development LOI may be found at the CTEP web site: http://ctep.cancer.gov/protocolDevelopment/letter_of_intent.htm#instructions.

7.2   The Drafting of a Protocol

A protocol is a clinical experiment's detailed written plan. This section details the essential features of a protocol. Careful attention to the following material will expedite CTEP's review of your protocol. For Cooperative Group studies that do not involve that a CTEP IND or supply of an agent by CTEP's Pharmaceutical Management Branch (PMB), the same essential features should be included in the protocol with appropriate adjustments.

7.2.1   Title Page

The protocol face sheet is the primary source of identifying information for the Protocol and Information Office (PIO) of CTEP, for the agent distribution system, for the IND file at the FDA, and for listing of the protocol in the Clinical Trials Reporting Program (CTRP, http://www.cancer.gov/ncictrp). Each protocol submitted to CTEP, therefore, must have a title page or face sheet that contains the following items:
- Version date of document
- Local protocol number (i.e., institution or group number) if applicable

- Title of study
- A single Protocol/Group Chair who will be responsible for interactions with CTEP for the study, including his or her name, institution/Cooperative Group, address, phone and fax numbers, and e-mail address (*A trainee may not be Protocol Chair-see* Section 12.2.2)
- Full name of institution/Group submitting the study
- List of each participating institution/Group/CCOP, and
- For Pharmaceutical Management Branch, CTEP, DCTD-supplied agents, a list of each agent by name and NSC number.
- For non-CTEP IND agents, list each agent by name with NSC number, IND number, and IND sponsor.
- Cooperative Groups may summarize by specifying "all Group members" or "restricted to..." and list institutions.
- All multicenter trials must include the CTEP Multicenter Guidelines which can be found on the CTEP web site at: http://ctep.cancer.gov/industryCollaborations/monitoring_multicenter.htm (see Section 7.2.17)

7.2.2   Schema

All treatment studies should include a brief schema depicting the treatment regimen(s).

7.2.3   Objective(s)

The objectives should be stated clearly, generally as hypotheses to be tested. The study design should be capable of answering the questions posed by the objectives. The statistical section should clearly state how the data will be analyzed in relation to each of the objectives. The hypotheses to be tested in ancillary studies also must be clearly stated, and the statistical section should address analyses of the data in relation to these hypotheses.

7.2.4   Background and Rationale

Sufficient background information should be included so that the study's rationale is clear. Unpublished data relevant to the rationale should be included in either this section, or, if extensive, as an appendix to the protocol. In addition to the background and rationale included for therapeutic aspects of a study, information should be provided to support ancillary studies to be performed. The rationale should be clearly stated for studying particular correlations between tumor characteristics and outcome measurements (response to therapy, disease-free survival, overall survival, etc.). The choice of the particular techniques to be used should also be justified.

7.2.5   Patient Eligibility Criteria

Patient eligibility criteria have been discussed previously (Section 4.1.3 and Section 5.2.2). Studies with objective response as an endpoint should include clear statements specifying whether tumor sites to be followed for response must be measurable, what criteria must be fulfilled to consider disease measurable, whether evaluable disease is permitted, and if so, at what sites. For ancillary studies, this section should include information regarding the choice of tumor sampling technique. For example, will aspiration biopsies be sufficient, or will surgical samples be required? How much tissue will be needed? What measures will be imposed to assure that the histopathologic diagnosis is not compromised? How will issues of tumor heterogeneity be addressed? What biases may be introduced by the sampling techniques and the amount of tissue required for the studies proposed?

7.2.6   Pharmaceutical Information

A separate pharmaceutical section is required for each agent. The content of the pharmaceutical section is dependent on whether the agent is investigational or commercial. A Pharmaceutical Data Sheet (PDS) is prepared by Pharmaceutical Management Branch (PMB) for most investigational agents. Regardless of whether the PMB data sheet is used, the following information about an agent is required in the protocol.

Investigational Agent Pharmaceutical Section
This section should include the following:
- **Agent Name and NSC number**
- **How Supplied**-Include the available dosage forms, ingredients, and packaging as appropriate. Also state the agent's supplier. For investigational agents sponsored by the Division of Cancer Treatment and Diagnosis, NCI, the supplier will be NCI and CTEP will have prepared a Pharmaceutical Data Sheet suitable for copying and pasting into the document as the pharmaceutical section for that agent.
- **Preparation** (how the dose is to be prepared)-Include reconstitution directions and directions for further dilution if appropriate.
- **Storage Requirements**-Include the requirements for the original dosage form, reconstituted solution and final diluted product, as applicable.
- **Stability**-Include the stability of the original dosage form, reconstituted solution and final diluted product, as applicable.
- **Route of Administration**-Include a description of the method to be used and the rate of administration if applicable. For example, continuous intravenous infusion over 24 hours, short intravenous infusion over 30 to 60 minutes, intravenous bolus, etc. Describe any precautions required for safe administration.
- **Unblinding Procedures**-If a protocol is a placebo-controlled blinded study, include language regarding the unblinding procedure. Please refer to Section 3.2.9.

Commercial Agent Pharmaceutical Section
- This section should include the following:
- Agent Name and NSC number
- **How Supplied:** State the agent's supplier, i.e., commercially available.
- **Preparation** (how the dose is to be prepared)**:** Investigators may refer the reader to the package insert for standard preparation instructions. If the agent is to be prepared in a 'non-standard' or protocol-specific fashion, the reconstitution directions and instructions for further dilution must be included. Appropriate storage and stability information should be included to support the method of preparation.
- **Route of administration:** Briefly describe how the agent will be administered in this protocol. For example, continuous intravenous infusion over 24 hours, short intravenous infusion over 30 to 60 minutes, intravenous bolus, etc.
- **Adverse Events:** The investigator may refer the reader to the agent's package insert. Note: The Informed Consent document should contain a list of all

expected adverse events that the patient is likely to experience. All adverse events in the informed consent should be written in laymen's terms.

7.2.7   Treatment Plan

Describe the protocol treatment clearly so all staff involved in the treatment of patients and in the conduct of the study can follow it. See Appendix IX, Guidelines for Treatment Regimens: Expression and Nomenclature.

7.2.8   Procedures for Patient Entry on Study

Procedures for patient entry, whether randomized or nonrandomized, should be specified. Required information includes a description of the randomization process and the patient characteristics and stratification factors (if any) to be provided at the time of entry. Patients must be registered on study prior to beginning treatment.

7.2.9   Adverse Events List and Reporting Requirements

Include a list and description of the reported adverse events and potential risks observed in pre-clinical and clinical studies. This information can be found in the Investigators Brochure for investigational agents or the package insert for commercial agents. For investigational agents supplied by CTEP, DCTD, the entire Comprehensive Adverse Events and Potential Risks list (CAEPR must be included in the body of the protocol. (If the CAEPR is included for an agent not supplied by PMB, the "SPEER" column may be deleted from the CAEPR as the SPEER information is specific to expedited reporting to CTEP.)

The CAEPR document is prepared using AE information from the Investigator Brochure, package insert, published papers, and submitted Adverse Event Expedited Reporting System (AdEERS) reports, and is typically updated yearly or more often as necessary. The CAEPR document consists of three sections.
The first section contains a comprehensive list (in table format) of reported and/or potential adverse events (AEs) associated with an agent using a uniform presentation of events by body system. A subset of the CAEPR document is a listing of Specific Protocol Exceptions to Expedited Reporting (SPEER) and appears in a separate column. The SPEER is a list of AEs that are protocol-specific exceptions to expedited reporting to NCI via AdEERS and should be reported only if they exceed the grade noted on this list. Refer to the "CTEP, NCI Guidelines: Adverse Event Reporting Requirements" http://ctep.cancer.gov/protocolDevelopment/electronic_applications/adeers.htm for further clarification. The second section of the CAEPR document lists events on agent trials where the relationship to the agent is undetermined. The third section of the CAEPR provides non-clinical toxicities (animal data) that have been reported for the agent; this third section is included only if there is insufficient human experience to assess safety events.

There are two types of CAEPR versions: 1.x (without frequencies) or 2.x (with frequencies). For version 2.x, the table of possible AEs is divided into three columns based on the reported frequencies: likely (>20%), less likely (≤20%), and rare but serious (<3%). Version 2.x is usually only prepared once AE data are available for ≥100 patients. CAEPRs utilize the latest Common Terminology Criteria for Adverse Events (CTCAE) version language when listing AEs and both the table of possibly related AEs and the list of reported but undetermined AEs, and are divided and organized by System

Organ Class (SOC). The possibly related AEs from the table in the first section of the CAEPR are used to generate the informed consent Risk Profile for an agent. The Condensed Risk Profile is a table of the possibly related adverse events for the specific agent described in laymen's terms, suitable for use in the informed consent. The Condensed Risk Profile is always distributed with the CAEPR. Instructions for reporting adverse events should be included in the protocol text, both for routine adverse event reporting as well as the expedited reporting of serious adverse events. (See <u>Section 11</u>)

7.2.10 Dose Modification for Adverse Events

The plan of dose change for adverse events should be stated for *each* study agent. Dose modification criteria should be described in NCI CTCAE terms.

Protocol Authors must carefully review all investigational agent Investigator's Brochures, as well as CTEP's "Comprehensive Adverse Events and Potential Risks" list (CAEPR) to be sure that they have included all reasonable measures to monitor expected adverse events. If the protocol includes an agent for which CTEP has prepared a CAEPR, it must be included whether CTEP is supplying the agent or not. These documents must be in the format provided and must be unaltered. (The investigator may delete the SPEER portion of the CAEPR if CTEP is not supplying the agent.)

7.2.11 Criteria for Response Assessment

The criteria for evaluating responses should be included. These should be specific for both measurable and evaluable disease. Disease-specific criteria are often required and should clearly indicate acceptable means of measurement, i.e., CAT scans, radio-nuclide scans, ultrasound, etc.

7.2.12 Monitoring of Patients

Specify how patients will be followed for assessment of treatment-related adverse events and therapeutic effect. A table of follow-up parameters that incorporates the schedule is particularly useful. The current version of CTCAE should be used when developing new DCTD-sponsored trials.

7.2.13 "Off-Study" Criteria

Criteria for terminating protocol treatment and/or removing a patient from treatment or from study should be specified.

7.2.14 Statistical Considerations

An adequate statistical section discusses the study design in relation to the objectives of the study and the plan for the evaluation of the data, specifically:
- Method of randomization and stratification
- Total sample size justified for adequate testing of primary and secondary hypotheses
- Error levels (alpha and beta) in phase 3 studies
- Differences to be detected for comparative studies
- Size of the confidence interval to be constructed around the estimated outcome
- Estimated accrual rate and/or study duration, with supporting documentation
- Stopping guidelines, including statistical and administrative procedures for monitoring the progress of the trial to implement early termination for very

positive results, or for results sufficiently negative to preclude the eventual achievement of statistically significant positive results
- Expected outcome parameters as appropriate (response rate, time to progression, survival times, etc.)
- Primary endpoint for interim and final analysis
- Clear specification of primary and secondary (e.g., subset) hypotheses
- Maximum number of patients
- Statistical analysis based on minority/gender.
- Plan for analysis

In accordance with the NIH guidelines on the inclusion of women and minorities as subjects in clinical research, the Department of Health and Human Services (HHS) requires that all pilot trials, phase 2 and phase 3 studies must include accrual targets for males, females and minorities. The accrual targets should reflect the expected accrual over the life of the study.

The policy states that women and members of minority groups and their sub-populations must be included in all NIH-supported biomedical and behavioral research projects involving human subjects, unless a clear and compelling rational and justification establishes inclusion is inappropriate with respect to the health of the subjects or the purpose of the research. The NCI suggests that the accrual targets be based on data from similar trials completed by your organization during the previous five years. It is hoped that the accrual targets will resemble the gender, ethnic and racial composition of the U.S. population as closely as possible. Please refer to the Protocol Submission Worksheet located on the CTEP website (http://ctep.cancer.gov/protocolDevelopment/default.htm) for permissible values and format requirements.

7.2.15 Records to be Kept
Specify the document on which each of the following is to be recorded, where it is to be sent, and on what schedule.
- On-study information, including patient eligibility data and patient history
- Flow sheets, or other forms for interim monitoring
- Specialty forms for pathology, radiation, or surgery when required, and
- Off-study summary sheet, including a final assessment by the treating physician
- What information will be recorded through remote data entry systems

7.2.16 Participation
All protocol treatments and observations will be made by investigator-physicians affiliated with a clinical site (refer to Section 3), and registered with CTEP. Under certain defined circumstances, it may be appropriate for interim treatments to be administered by certain physicians not registered with CTEP (other than trainees, who are assumed to be under the supervision of a registered investigator). In such cases, the protocol should state:
- Precisely what responsibilities those physicians will assume, including response assessment, and adverse event reporting
- How dose modifications will be decided and reported, and the mechanism by which data needed for evaluating adverse events and response will be transmitted to the registered investigator responsible for the patient, and

- The intervals at which a patient must be evaluated by an investigator-physician at the clinical site

See Section 14 for further details concerning which physicians may actively participate in a clinical trial involving CTEP-supplied investigational agents.

7.2.17 Multicenter Trials

CTEP Multicenter Guidelines
If an institution wishes to collaborate with other participating institutions in performing a CTEP sponsored research protocol, then the following guidelines must be followed. CTEP Multicenter Guidelines can be found on the CTEP web site at: http://ctep.cancer.gov/industryCollaborations/monitoring_multicenter.htm.

Cooperative Group studies follow the "Cooperative Group Guidelines" available on the CTEP website under Guidelines/Policies section at: http://ctep.cancer.gov/investigatorResources/default.htm#guidelines_policies.

Responsibilities of the Protocol Chair:
- The Protocol Chair will be the single liaison with the CTEP Protocol and Information Office (PIO) and is responsible for
  - coordinating, developing, submitting, and receiving approval for the protocol and subsequent amendments.
  - rewriting or modifying the protocol
  - assuring that all participating institutions are using and identical, current version of the protocol.
- The Protocol Chair is responsible for the overall conduct of the study at all participating institutions and for monitoring its progress. The Protocol Chair must assure that all reporting requirements to CTEP are properly adhered to.

AdEERS is programmed for automatic electronic distribution of reports to the following individuals: Study Coordinator of the Lead Organization, Principal Investigator, and the local treating physician. AdEERS provides a copy feature for other e-mail recipients.

- The Protocol Chair is responsible for the timely review of Adverse Events (AE) to assure safety of the patients.
- The Protocol Chair will be responsible for the review of and timely submission of data for study analysis.

Responsibilities of the Coordinating Center:
Each participating institution will have an appropriate Federal Wide Assurance (FWA) on file with the Office for Human Research Protections (OHRP), HHS. The Coordinating Center is responsible for assuring that each participating institution has a valid up to date FWA and must maintain copies of IRB approvals from each participating site. Prior to protocol activation at each participating institution, the coordinating center must submit documentation of initial IRB approval to the CTEP PIO. (Please Note: Cooperative Group and Consortium clinical sites are to submit IRB approvals through the CTSU Regulatory Support System accessed at https://www.ctsu.org/public/rss2_page.aspx).

The Coordinating Center is responsible for
- assuring that each participating site has obtained IRB approval before the site registers its first patient

- ensuring each participating site is accruing a representative sample consistent with the estimate of population representation in the site's geographical location for race and ethnic groups. Refer to the Gender and Minority Accrual Estimates section of the protocol submission worksheet (http://ctep.cancer.gov/protocolDevelopment/docs/psw.pdf).
- preparing all submitted data for review by the Protocol Chair

- maintaining documentation of AE reports using one of two options for AE reporting:
    - (1) participating institutions may report directly to CTEP with a copy to the Coordinating Center; or,
    - (2) participating institutions report to the Coordinating Center who in turn reports to CTEP. The Coordinating Center will submit AE reports to the Protocol Chair for timely review.

Audits may be accomplished in one of two ways (1) source documents and research records for selected patients are brought from participating sites to the Coordinating Center for audit; or, (2) selected patient records may be audited on-site at participating sites. If the NCI chooses to have an audit at the Coordinating Center, then the Coordinating Center is responsible for having all source documents, research records, all IRB approval documents, NCI Investigational Agent Accountability Record forms (DARF), patient registration lists, response assessments scans, x-rays, etc. available for the audit.

Inclusion of Multicenter Guidelines in the Protocol:
The protocol must include the following minimum information:
- The title page must include the name and address of each participating institution and the name, telephone number and e-mail address of the responsible investigator at each participating institution.
- The Coordinating Center must be designated on the title page.
- Central registration of patients is required. The procedures for registration must be stated in the protocol.
- Data collection forms should be of a common format. Sample forms should be submitted with the protocol or the protocol must state that a remote data capture system will be utilized. The frequency and timing of data submission forms to the Coordinating Center should be stated.
- Describe how AEs will be reported from the participating institutions, either directly to CTEP or through the Coordinating Center.

Drug Ordering:
Except in very unusual circumstances, each participating institution will order DCTD-supplied study agents directly from CTEP. Study agents may be ordered by a participating site only after the initial IRB approval for the site has been forwarded by the Coordinating Center to the CTEP PIO or to the CTSU for those protocols where they are providing regulatory support.

7.3   Informed Consent

Each informed consent document must be protocol-specific and follow the current NCI Informed Consent Template. The current Template may be found on the CTEP website (http://ctep.cancer.gov/protocolDevelopment/default.htm#informed_consent) or the NCI website at the section titled *Recommendations for the Development of Informed Consent Documents for Cancer Clinical Trials* (http://www.cancer.gov/clinicaltrials/understanding/simplification-of-informed-consent-docs/allpages). Use of the Template will ensure that all of the elements required by Federal regulation (21 CFR 50.25 and 45 CFR 46.116) are included while providing the required information in a concise manner. CTEP will not approve a protocol if its informed consent form fails to follow the current NCI Informed Consent Template or comply with the Federally-required elements.

Individual institutions may make minor changes to model informed consent forms. However, the informed consent document's originator must approve any changes in risks or alternative procedures.

CTEP has developed a Condensed Risk List for each of its IND agents and several commonly used commercially available chemotherapy drugs. The appropriate List is provided to the Protocol Chair at the time of LOI/Concept approval. The List may be copied-and-pasted into the 'Risks' section of the informed consent form as applicable for the study.

CTEP uses the CAEPR, which uses CTCAE terminology, to determine what possible side effects should be included in each agent's Risk List. Each CTCAE term has been translated into a lay term for use in the Risk List. A "lay translation spreadsheet" of CTCAE Terms and the respective lay term for each is available on the CTEP website.

If a Condensed Risk List is required by the study for a drug not yet available on the CTEP website, protocol authors can custom-build a list. The following documents are posted on the CTEP website to assist in development of custom-built lists:
> Instructions for building Condensed Risk Lists for Informed Consent Forms
> CTEP CTCAE Term-LayTerm Translation Spreadsheet

Custom-building a Condensed Risk List as instructed will ensure that the description of expected adverse events is complete, consistent, and reflective of the treatment plan to be used. Consult the Investigator's Brochure, or other comprehensive resource, for information about expected adverse events to include in the list. Possible side effects pertaining to all treatment modalities involved in the research must be described.

Please refer to Appendix V for links to guidance resources for informed consent.

7.4   Protocol Templates

CTEP has developed several model protocol templates. To facilitate rapid review, you are encouraged to use the NCI protocol templates which can be found on the CTEP web site. (http://ctep.cancer.gov/protocolDevelopment/templates_applications.htm). An agent specific protocol template may be available for a particular investigational agent and will accompany the LOI/Concept approval letter or may be requested from the PIO at PIO@ctep.nci.nih.gov.

8.1   The Protocol and Information Office (PIO)

Within CTEP, the PIO manages the review process and maintains the official record of all CTEP-supported protocols, amendments, and protocol-related communications. PIO is also involved with processing all protocol submissions to the FDA. The PIO maintains more than 1000 active protocols.

All protocols and related correspondence should be sent directly to the PIO via email at pio@ctep.nci.nih.gov. *The PIO will distribute all email to the appropriate CTEP physician staff.* Please do not direct protocol-related materials to any other CTEP staff member as this will lengthen the time required for resolution or review.

When submitting each new protocol, please also include a completed copy of the current version of the Protocol Submission Worksheet, http://ctep.cancer.gov/protocolDevelopment/default.htm. Direct all telephone calls regarding status of protocol and amendment reviews to the PIO at 301-496-1367.

8.2   How to Submit a Protocol

Send each new protocol directly to the PIO and include:
- The electronic copy of the Protocol Submission Worksheet,
- An electronic copy of a paginated, legible protocol, including a local protocol number. Please be certain the protocol document contains information about *each* of the topics listed in Section 7.2 and Section 7.3.

If these items are missing or incomplete, PIO will place the submission on hold, contact the submitter and suspend further processing or review until the submission is complete.

Upon receipt, PIO assigns an NCI protocol number to each protocol, and acknowledges your protocol submission with the NCI-assigned protocol number. *You must reference this NCI protocol number in all subsequent communications with CTEP regarding this study.* Assignment of this number does not imply approval; only the final approval letter signifies approval and the authorization to order study agents.

8.3   IRB Approval

Each investigator must meet the requirements of the Federal regulations for Protection of Human Subjects, and Informed Consent, and IRB Review and Approval (45 CFR 46: http://www.access.gpo.gov/nara/cfr/waisidx_01/45cfr46_01.html and 21 CFR 50 and 56: http://www.access.gpo.gov/nara/cfr/waisidx_01/21cfr50_01.html. Also see Section 12.1.1 and Section 12.1.2). Each protocol must have documentation of IRB approval prior to CTEP approval.

Cooperative Groups, CTEP funded Consortium protocols and other institutions utilizing the CTSU Regulatory Support System to meet IRB approval documentation requirements may find the most current information on the following CTSU website https://www.ctsu.org/public/rss2_page.aspx.

IRB approval for a single institution study is required in order to obtain CTEP approval and activate a protocol. The IRB approval documentation must refer to the same protocol version as the most recent protocol version submitted to CTEP for approval.

For multi-center non-Group studies, the Coordinating Center and each participating site must submit documentation of IRB approval. All IRB approvals are submitted to the PIO via the lead institution. Upon receipt of the first IRB approval from any of the participating sites, the protocol is evaluated for approval. Each site undergoes individual activation upon receipt of IRB approval. The IRB approval documentation submitted for activation must document the current active version of the protocol. Since sites activate on a rolling basis, each site should verify the current active version of the protocol with the Coordinating Center. Activation of a protocol is required before study agent can be distributed by the Pharmaceutical Management Branch to the site. Submission of IRB approvals for subsequent protocol amendments is not required.

Investigators' failure to submit evidence of IRB approval is a major cause of delay in protocol approval. The documentation requirements are satisfied by completion of Form 310, or by a letter specifying:
- The study title and document version date
- The site's Principal Investigator and the institution's assurance number; i.e., the assurance number issued by the OHRP, DHHS
- Date of IRB review
- Date of approval's expiration
- The dated signature of an institutional official, usually the IRB Chair.

8.4   Protocol Review

CTEP must review and approve every protocol involving CTEP-supplied study agents or studies receiving NCI support or funding. CTEP reviews each protocol for completeness, scientific merit, duplication of existing studies, patient safety, and adequacy of regulatory and human subjects' protective aspects.

CTEP will inform the protocol source if the protocol is incomplete, or the investigator/institution is ineligible under the proposed category of sponsorship. CTEP will not review the study for scientific content under these circumstances.

8.4.1   Scheduling the Protocol for Review

The PIO schedules the protocol for review by the Protocol Review Committee (PRC). The cutoff date for protocol receipt is Tuesday at 5 pm. Protocols that are complete and eligible are abstracted into the CTEP system and scheduled for review within four weeks. Please see Section 8.2 for a discussion on the handling of incomplete submissions.

8.4.2   Review by the CTEP Protocol Review Committee

The CTEP PRC reviews all protocols, except those that are the purview of disease-specific Steering Committees (see Sections 6.2 and 6.4). This committee, composed of CTEP's professional staff and consultants from other NCI divisions, and chaired by the Associate Director, CTEP, meets weekly and usually reviews 10 to 20 documents (protocols, LOIs and concepts) at each session.

Each protocol is assigned a minimum of five reviewers; as many as six to seven may be required for complex multimodality protocols. The protocol and informed consent form are reviewed by an oncologist(s), biostatistician, pharmacist and regulatory affairs professional(s) with expertise in informed consent issues.

8.4.3   The Review of the Protocol

The PRC discusses the protocol after hearing the reviews of each assigned reviewer and makes a decision that the science and safety of the study are:
Approved as written;

- *Approved with recommendations* - CTEP asks the investigator to consider points raised in the consensus review, but the investigator is not obligated to revise the study. If changes are made prior to activation of the study, the investigator must send CTEP an amendment prior to activation that details any changes in the CTEP-approved document;
- *Approval deferred pending revisions* - The PRC has significant questions about the proposed study. It cannot be accepted unless the investigators satisfactorily address the concerns of the written consensus review. The investigator should submit a revised protocol within two weeks of receipt of the consensus review (Section 8.4.7);
- *Disapproved* - In the judgment of the PRC, the protocol cannot be approved even with major revisions.

8.4.4   The Review of the Informed Consent

The PRC also reviews the informed consent document to be certain that:

- The document includes all required elements of informed consent mandated by Federal regulation; and
- The description of potential benefits and adverse events is complete and accurate.

Investigators should communicate changes made to the consent document resulting from a CTEP review to their IRBs. It is our intent to see that the informed consent complies with the NCI template.

It is not CTEP's intent for their informed consent review to supplant IRB review. Provided the consent document meets the requirements of regulation and law and contains sufficient information to enable an individual to make an informed choice, the local IRB approval of the content is generally to be regarded as definitive.

Individual institutions may make minor changes to the CTEP-reviewed informed consent form (http://ctep.cancer.gov/forms/default.htm). However, by the originator of the informed consent document must approve any changes in risks or alternative procedures.

8.4.5   The Review of Regulatory and Administrative Concerns

The PRC reviews each protocol to assure proper instructions for reporting adverse events are included, an accurate and up-to-date pharmaceutical section is provided, and necessary instructions for multicenter trials are given, if appropriate

8.4.6   The Review by the Pharmaceutical Collaborator

In cases where a Pharmaceutical Collaborator is involved, CTEP forwards protocols received to the Collaborator for review and comment approximately two weeks before it is reviewed by the PRC. CTEP discusses Collaborator comments if they are received before the Protocol Review Committee meeting. They give them due consideration, and comments from either Collaborator or the CTEP staff that are agreed upon in the PRC meeting are included in the consensus review.

8.4.7   The Consensus Review

After the PRC meeting, the primary reviewer generates a consensus review that states the PRC's collective concerns. PIO sends this consensus review with a cover letter stating the summary PRC decision to the protocol originator within approximately 30 days of receipt of a complete protocol.

8.4.8   Responding to CTEP Consensus Review

If the protocol or the informed consent requires revisions, the investigator should send a revised submission to the PIO. A complete submission includes the revised protocol and informed consent with a cover letter that details the responses to the points raised in the consensus review. If CTEP reviewers find the response satisfactory, then the protocol goes forward for final approval.

CTEP's consensus reviewer may choose to send the revised protocol back to the full PRC for further consideration. If the PRC still does not accept the protocol, CTEP sends a letter to the investigator detailing remaining CTEP concerns. The investigator should respond to this re-review in the same way as described for the initial consensus review. This process continues until the science, regulatory, and administrative aspects, and informed consent are each accepted or the study is withdrawn or disapproved. Each evaluation of a revised study requires an average of 10 to 30 days.

Under the new Operational Efficiency Working Group (OEWG) timelines and processes, it is anticipated that CTEP and CTEP-sponsored investigators will work closely in a collaborative fashion to resolve any pending issues quickly. Via conference calls and other interactive media, the expectation is that all parties contributing to the development of the protocol will work together to achieve the target timelines for activation of phase 1, 2, and 3 trials. It is hoped that the number of protocol revisions will be held to 1 or 2 revisions at most in order to achieve timely activation.

8.5   Protocol Document Approval

Upon final acceptance of the protocol and the informed consent, as indicated by the concurrence of the consensus reviewer and appropriate CTEP staff, and once all regulatory, and agent supply issues are in order, CTEP sends a letter of protocol approval the protocol originator.

Please note that protocol approval will not be given by telephone. Although CTEP staff may discuss the study with Protocol/Group Chairs, they should consider nothing official until they receive written notice via email. After PIO sends written approval, PMB will accept orders for study agents. All approved protocols using DCTD-sponsored agents are submitted to the FDA as part of the IND file.

Cooperative Groups and CTEP-funded Consortia provide evidence of protocol and informed consent by Institutional Review Boards (IRBs) in a separate, but similar process, as these multi-institutional clinical trial networks do not allow patient enrollment unless they receive confirmation of IRB approval at their Operational Centers.

8.6   Amendments

8.6.1   Protocol Amendments

Any change to the approved protocol document must be documented point by point in a cover letter, and a replacement page(s) or a revised protocol document and informed consent submitted to the PIO. Please reference the NCI protocol number, date each amendment, and document version number sequentially for each study. Upon receipt, CTEP staff reviews each amendment. Further detailed guidance on amendment submissions can be found in the CTEP Amendment Request Submission Policy: http://ctep.cancer.gov/protocolDevelopment/default.htm.

8.6.2   Amendments Initiated by CTEP

CTEP issues 3 general types of communication regarding patient safety and amendment requests for clinical trials:

1. Safety Notices: Safety Notices are sent when an adverse event report is filed with the FDA. Safety Notices are provided for informational purposes and CTEP does NOT require any action. Institutional Review Boards (IRBs) have discretion regarding actions they take regarding these notices.

2. General Requests for Amendments: General requests for amendments are not specific to patient safety issues. A "Request for Amendment" is also used for other types of protocol modifications such as a change in an agent's formulation or the roll-out of a new CTCAE version. Safety-related "Requests for Amendment" will generally be used if a CAEPR is updated and may precede an Action Letter if the new risks are significant.

3. Action Letters: Action Letters require some action from an Operations Center or Principal Investigator. The timing and nature of the action (i.e., amendment, accrual closure) is dependent on the type of Action Letter. These types are based on the impact of the information and requested action (e.g., inclusion of a new risk that affects a patient's risk/benefit in a specific trial). Failure to respond in a timely fashion may result in temporary study closure or PI suspension.

8.6.3   Response to FDA Request for Information

Upon the review of a protocol the FDA may forward comments or recommendations via CTEP. The receipt of FDA comments from the Regulatory Affairs Branch, CTEP, should be acknowledged by e-mailing: RABLetters@tech-res.com. A site that receives an FDA Request for Information communication from CTEP should respond to each point and if a response results in a change in the protocol, a protocol revision or amendment must be submitted along with your response to CTEP. A response to FDA must be identified on your change memo as "Response to FDA Request for Information", preferably near the top of the letter or memo; and bolded, when submitted to the Protocol and Information Office (PIO@ctep.nci.nih.gov). The timelines for response is specified in the communication to the site and is generally either 30 or 45 days.

8.7   Study Status

Investigators should communicate changes in study status to the CTEP PIO immediately. The following list shows the categories of study status recognized by CTEP and PDQ:

- AP (Approved) – Protocol has received final CTEP approval.
- AC (Active) – Trial is open to accrual.
- TC (Temporarily Closed to Accrual) – Protocol is temporarily not accruing.
- TB (Temporarily Closed to Accrual and Treatment) – Protocol is temporarily not accruing and patients are not receiving therapy.
- CA (Closed to Accrual, Patients still on Treatment) – The protocol has been closed to patient accrual. Patients are still receiving therapy.
- CB (Closed to Accrual, All Patients have Completed Treatment) – The protocol has been closed to patient accrual. All patients have completed therapy, but patients are still being followed according to the primary objectives of the study. No additional study agents are needed for this study.
- CP (Completed) – The protocol has been closed to accrual, all patients have completed therapy, and the study has met its primary objectives. A final study report/publication is attached or has been submitted to CTEP.
- AD (Administratively Completed) – The protocol has been completed prematurely (e.g., due to poor accrual, insufficient agent supply, IND closure). The trial is closed to further accrual, and all patients have completed protocol treatment. A final study report (see below) is not anticipated.

8.8   Reactivation of Studies

Protocols that are temporarily closed require written CTEP approval before reactivation if they are:

- Temporarily closed for reasons of patient safety or regulatory issues; or
- Closed for reasons of peer review, site visit, or other NCI-initiated reasons.

Protocols that are temporarily closed to accrual by the Lead Organization based on early stopping rules do not require CTEP approval for reactivation.

The DCTD/CTEP will provide study agents to registered investigator-physicians for use in:
- CTEP-approved protocols,
- Special Exception (see Section 17.1) and
- Treatment Referral Center (TRC) requests (see Section 17.2)

NCI registered investigators must have the following documents on file with the Pharmaceutical Management Branch (PMB) (see Section 12.1).
- Current FDA 1572,
- Supplemental Investigator Data
- Financial Disclosure Forms, and
- Current CV

These forms are available at http://ctep.cancer.gov/forms/index.html.
Unless the protocol is conducted by an NCI Cooperative Group or Consortium that maintains their investigator roster through the NCI Regulatory Support System, all investigators eligible to receive agents must be named on the protocol title page. Sites will only use study agents supplied by DCTD to treat patients entered onto the CTEP-approved protocol for which the DCTD has agreed to supply agent.

9.1   How to Place Agent Orders

Active CTEP registered investigators or their authorized shipping designees and ordering designees may order agents from the Pharmaceutical Management Branch (PMB), CTEP for NCI- sponsored or funded clinical trials using PMB-supplied agents.

Active CTEP-registered investigators and investigator-designated shipping designees and ordering designees can submit agent requests through the PMB Online Agent Order Processing (OAOP) application, https://eapps-ctep.nci.nih.gov/OAOP/pages/login.jspx. Access to OAOP requires the establishment of a CTEP Identity and Access Management (IAM) account, https://eapps-ctep.nci.nih.gov/iam/, and the maintenance of an "active" account status and a "current" password. The OAOP application will prompt users for all necessary information needed to submit the agent requests and perform all validations as data is entered.

9.2   Particular Points to Note

Routine Orders: Normal PMB processing time is two (2) working days. PMB will ship orders within two working days, based on the agent's availability and provided there are no shipping restrictions (e.g. holidays or thermo-labile agents that are shipped Monday through Thursday only). Sites can make special arrangements for Saturday delivery by contacting PMB. Refer to individual protocols for specific ordering instructions.   Orders are shipped by U.S. Postal Service Priority Mail or other ground service. Generally allow up to one week for order delivery. Agents having special storage conditions [e.g., thermo-labile agents that require refrigerated or frozen (-20 or -70 degree C) storage conditions] or shipping requirements (e.g., dangerous goods, infectious substances) are shipped Monday through Thursday for next day delivery.
- Urgent Orders: PMB provides next day delivery to registered investigators to meet "emergency" or urgent needs. Requests for next day delivery must be received at PMB by 2:00 p.m., Eastern Time. The requirement for next day

delivery must be stated on the order request and an express courier account number provided.

- When a number of investigators are participating on a clinical study at the same institution, one investigator should be considered or designated the principal or lead investigator under whom all investigational agents for that protocol should be ordered.
- Orders will only be shipped to the investigator's designated shipping address. Investigators may only have a single shipping address.
- All changes to the investigator's shipping address or changes or additions to investigator-designated shipping designees and ordering designees must be in writing by submission of an updated Supplemental Investigator Data Form signed by the investigator or by submission of a Primary Shipping Address and Designee (PSD) request form or PSD update form.
- For OAOP-submitted orders, submitters will receive a confirmation e-mail of successful order submission. Order status may be viewed at anytime through the OAOP application. Upon shipment, a confirmation e-mail is sent which includes the order details and tracking information (if appropriate).
- Limit agent requests to an eight week supply.
- Avoid ordering excessive quantities. CTEP will reduce the quantity shipped if the order is excessive in relation to protocol requirements, or if CTEP's inventory is insufficient at that time.
- Due to added quality control steps taken to protect the blind, next day delivery is not available for blinded study materials.

9.3   Routing of Agent Requests

- Investigators should submit agent requests directly to the Pharmaceutical Management Branch.
- Blinded Studies: For most blinded clinical studies, agent requests are submitted electronically to PMB by the lead organization at the time of patient registration and randomization. Check the protocol for procedures for agent requests.

9.4   Affiliates and Study Agent Orders

- Investigators at affiliated institutions or clinics should order CTEP-supplied study agents directly from PMB. CTEP-supplied study agents must not be repackaged and shipped or forwarded to an affiliate or patient. PMB's policy of shipping study agents to the investigator's institution or practice site assures that all investigators receiving study agents are registered with PMB; simplifies agent tracking and accountability; minimizes correspondence delays in emergencies; assures agent integrity; reduces administrative workload; and eliminates secondary shipping expenses.
- PMB policy allows centralized pharmacies to receive study agents for re-distribution to local satellite institutions and affiliated investigators who are registered with PMB and have designated such a "central pharmacy" as their shipping address. Sites should coordinate such arrangements with the PMB.
- The central pharmacy must ensure that all investigators receiving study agents have a current FDA Form 1572 on file with PMB.
- Satellites must maintain dispensing records.
- Local satellite institutions or affiliates must be serviced by bonded institutional couriers or the study staff.

- CTEP-supplied study agents must never be repackaged or forwarded by mail or express courier.
- Institutions that are separated geographically, requiring study agents to be mailed, are not considered satellites for agent accountability purposes and should receive study agents directly from PMB.
    - These sites may order directly or in some situations, the central pharmacy may submit orders to PMB for delivery to participating investigators.
    - CTEP policy forbids secondary distribution of study agents to other physicians, or transfer of study agents between institutions. CTEP intends that study agents be distributed directly to investigators. If under exceptional circumstances emergency transfer seems justified, explicit pre-approval by the Pharmaceutical Management Branch is required.
    - If investigators wish to use a local oncologist to administer some treatments for late phase 2 or phase 3 trials, they must contact the Pharmaceutical Management Branch for assistance. In such situations, the local oncologist(s) must register with PMB and be covered by an appropriate IRB. PMB will ship the agents directly to the local investigator with the enrolling physician's approval.
    - CTEP policy does not allow shipment of agent directly to patients. Sites may not re-ship agent to other sites or to patients.
    - Finally, CTEP may approve a special distribution arrangement for certain unusual circumstances.

9.5   Requests for Non-research (Treatment) Use of Investigational Agents

Requests for use of DCTD investigational agents under TRC or Special Exception categories must satisfy certain requirements. These considerations are detailed in Section 17.1 and Section 17.2 of this manual.

9.6   Requests for Non-clinical (Laboratory) Use of Investigational Agents

Clinical trials investigators cannot take investigational agents for non-clinical or laboratory use from supplies received for CTEP supported clinical trials. Interested investigators should direct their agent requests for non-clinical use of investigational agents for preclinical or laboratory experiments to the Regulatory Affairs Branch (RAB), NciCtepRequests@mail.nih.gov.

A non-clinical request form will be provided for detailed information about your proposal. RAB, in conjunction with the IDB drug monitor and the pharmaceutical collaborator, will evaluate your request. If it is approved, RAB will execute a Materials Transfer Agreement.

9.7   Status of Investigational Agents Following FDA Marketing Approval

In an attempt to identify more effective therapeutic regimens, the DCTD continues to evaluate agents in ongoing clinical trials after they have received FDA marketing approval for a particular cancer indication. Physicians who have initially registered a patient to receive an agent under a TRC or Special Exception protocol before an agent is approved by FDA may continue to receive the agent at no cost from CTEP for the registered patient. This is not a firm policy; it depends on the cooperation of the manufacturer (sponsor) supplying the agent.

Commercially-available agents supplied to investigators or physicians under an NCI protocol, including a TRC, or Special Exception protocol, are still considered study agents. Therefore, the same accountability requirements for study agents in place prior to the agent's approval continue to apply.

10.1   Introduction

Investigators who test IND study agents have an important responsibility: timely, accurate reporting of data to the trial sponsor. Prompt provision of these reports is not an arbitrary requirement. The information contained in them informs CTEP about the agent's development progress, sometimes suggesting promising new directions. In addition, CTEP requires investigators to meet obligations under FDA regulations to

(a) monitor the study and

(b) submit reports of current findings

Failure to comply with these reporting requirements is a serious breach of the agreement that each investigator makes in signing the FDA Form 1572, http://ctep.cancer.gov/forms/index.html, and may result in suspension or termination of investigator privileges.

For all trials, investigators must report two types of data: individual patient data and study summaries. Each is briefly discussed below. Then, Section 10.2.1 details specific reporting requirements for phase 1 trials, and Section 10.2.2 details the same for phase 2 and 3 trials.

Case Report Forms - Information about patients is recorded on case report forms or through Remote Data Entry (RDE) (or in a computerized clinical trials database) that incorporate all patient data stipulated in the protocol.

Data submitted via RDE is different than patient source documentation, usually the patient's medical chart. The study specific case report form or computerized protocol electronic data serves as the formal and fixed data base on which the study is reported. The patient's primary medical record is generally not organized for clinical research and does not reliably contain the treating physician's assessment of the effects of the protocol treatment. For these reasons, investigators and their staff should maintain a separate research record (i.e., case report form or well-designed clinical trials database) on each protocol patient in a real time manner.

A research record should also include the responsible physician's assessment of the treatment effect (e.g., response category) and judgment as to whether any medical events in the patient's course were treatment-induced (i.e., agent-related adverse events).

Procedures for reporting study data vary according to the type of study and category of sponsorship. They are outlined below for the three phases of clinical agent development. Adverse events are reported through the Adverse Event Expedited Reporting System (AdEERS) using the Common Terminology Criteria for Adverse Events (CTCAE) and is discussed in Section 11.

10.2   Report Requirements of CTEP Supported Trials

10.2.1 Clinical Trials Monitoring Service (CTMS)

Clinical Trials Monitoring Service (CTMS) reporting is required for all early phase 1 studies and some selected phase 2 trials. The criteria to determine if a study is early phase 1 include the first time the agent is used in human studies or the first time a new agent combination is used in humans or the first time an agent or combination is used in a specific patient group (e.g., children). Concerns regarding the frequency or severity of adverse events may also warrant assignment to CTMS monitoring. CTEP staff will advise the site on the reporting mechanism to use for each trial.

For each patient on trial, investigators must maintain data in an electronic format. All information specified in the protocol should be recorded and must be maintained in real-time. The site submits these forms electronically on a biweekly basis to the CTMS. The biweekly submission includes case report updates on patients actively on study and data on all new patients entered since the last submission.

These reporting requirements apply to *all* phase 1 trials of agents newly entering clinical trials, including adult and pediatric studies. The Protocol Chair will receive a full report of his or her study from the CTMS. The CTMS provides CTEP with summary information on all trials in the database for each agent. The CTMS analyzes the data from phase 1 trials for timeliness of submission and completeness and provides monthly reports of this analysis to CTEP. Additionaly, Princial Investigators receive quarterly reports from CTMS. The CTMS data is also loaded monthly into the Clinical Data Update System (CDUS).

10.2.2 Clinical Data Update System (CDUS)

The quarterly Clinical Data Update System (CDUS) is the primary clinical trial data resource for DCTD and the Division of Cancer Prevention (DCP).

This includes all:
- DCTD/DCP-sponsored Cooperative Group and Community Clinical Oncology Program (CCOP) Research Base treatment trials using DCTD-supplied study agents;
- DCTD/DCP- sponsored, supported or funded Cooperative Group and CCOP Research Base treatment trials using non-NCI agents;
- DCTD/DCP-grant funded non-Cooperative Group (Cancer Center or other institution) trials (if CDUS reporting is a grant requirement) using non-NCI agents;
- DCTD/DCP-sponsored Cooperative Group and CCOP Research Base non-treatment trials (accrual >100 patients.); and
- DCP-sponsored CCOP Research Base cancer prevention and control trials.

Investigators should refer to their individual document for guidance on the specific data reporting system required.

Abbreviated CDUS
The abbreviated CDUS requires quarterly submission of protocol administrative information (e.g., status) and patient-specific demographic data (e.g., gender, date of birth, race, etc.).

Complete CDUS

The complete CDUS data set includes information obtained in the abbreviated CDUS data set as well as patient administrative information (e.g., registering institution code, patient treatment status), treatment information (e.g., agent administered, total dose per course), adverse event (AE type and grade), and response information (e.g., response observed, date response observed).

A complete discussion of the CDUS reporting requirements is found in the CDUS Instructions and Guidelines on the CTEP web site at the following address: http://ctep.cancer.gov/protocolDevelopment/electronic_applications/cdus.htm.

10.2.3 Adverse Events

The importance of reporting adverse events (AE) as described above cannot be overstated. In addition, Section 11 discusses *types of events that must be reported in an expedited manner to CTEP. These serious adverse events are submitted via the Adverse Event Expedited Reporting System (AdEERS).* (http://ctep.cancer.gov/protocolDevelopment/electronic_applications/adeers.htm)

10.2.4 Study Status

The Principal Investigator or their designee must communicate study status changes promptly in writing to the CTEP PIO using the Protocol Status Update Form (See Section 8.7). Telephone discussions with CTEP physician staff are not considered formal notice of status changes.

10.2.5 Amendments

All protocol amendments must be submitted to and approved by CTEP prior to implementation (see Section 8.6).

10.2.6 Publications for Studies Conducted Under a CTEP Collaborative Agreement

Investigators must provide all manuscripts reporting the results of DCTD-supported clinical trials conducted under a CTEP Collaborative Agreement to CTEP for immediate delivery to the pharmaceutical collaborator for advisory review and comment prior to submission for publication. The pharmaceutical collaborator will have 30 days from the date of receipt for review. An additional 30 days may be requested by the pharmaceutical collaborator in order to ensure that confidential and proprietary data, in addition to the company's intellectual property rights, are protected.

Abstracts to be submitted by investigators must be sent to CTEP for forwarding to the pharmaceutical collaborator for courtesy review with sufficient time to allow the collaborator at least three days for review. Before submitting a manuscript for publication or otherwise publicly disclosing information (i.e., abstracts, posters, presentation slides) concerning an agent under an NCI collaborative agreement. Investigators must send any such disclosure, marked CONFIDENTIAL, to the Associate Chief, Regulatory Affairs Branch (RAB), electronically at NCICTEPpubs@mail.nih.gov. RAB will coordinate and expedite the pharmaceutical collaborator's review.

Investigators must send any publication resulting from a DCTD-sponsored study to the Protocol and Information Office, CTEP, (PIO@ctep.nci.nih.gov) identifying the protocol by the NCI protocol number and grant/contract number.

10.2.7 Inventions

Clinical investigators are required to report any inventions relating to studies under an NCI-funded protocol, before public disclosure to National Institutes of Health (NIH), Division of Extramural Inventions and Technology Resources (DEITR), OPERA, Office of Extramural Research (OER), 6705 Rockledge Drive, Suite 310, MSC 7980, Bethesda, MD 20892-7980 Attn: Director, DEITR. The report should be in sufficient detail so as to enable the government to evaluate any potential contributions in the technological advances by NCI scientists. To allow more effective evaluation, the NCI asks that the investigator send an additional copy of the invention report to: Associate Chief, Regulatory Affairs Branch, Cancer Therapy Evaluation Program, DCTD, NCI, Executive Plaza North, Suite 7111, Bethesda, MD 20892 [NCICTEPpubs@mail.nih.gov].

Information regarding the intellectual property options agreed to by the Investigator's Institution to the pharmaceutical collaborator can be found at http://ctep.cancer.gov/industryCollaborations2/intellectual_property.htm.

10.3   Record Retention

FDA regulations require investigator to keep all research records (including patient charts, case report forms, x-rays and scans that document response, IRB approvals, signed informed consent documents and all agent accountability records) for at least 2 years after an NDA or BLA has been approved for that indication, or the CTEP,DCTD IND has been withdrawn from the FDA. CTEP will notify investigators when these events occur. This requirement is an explicit part of the FDA Form 1572, http://ctep.cancer.gov/forms/.

10.4   Reporting to IRBs

All studies supported in any way by CTEP must be under the auspices of an IRB that has obtained an approved Federal Wide Assurance (FWA) issued by the Office for Human Research Protections (OHRP), HHS.

Each investigator must report any problems, serious adverse events, or proposed changes in the protocol that may affect the investigation's status patients' willingness to participate in it to his or her IRB. At intervals appropriate to the study's degree of risk, the investigator must report to the IRB. These intervals must be no less frequent than once a year, and when the study is complete.

11   Adverse Events

Because many anticancer agents have narrow therapeutic indexes, adverse events (AEs) commonly accompany treatment. In addition, cancer patients often exhibit signs and symptoms attributable to cancer or its complications. For these reasons, a medical event's definition and identification as an AE related to an anticancer agent presents a special challenge for the investigator and IND sponsor.

The most current version of the NCI Common Terminology Criteria for Adverse Events (CTCAE) (http://ctep.cancer.gov/protocolDevelopment/electronic_applications/ctc.htm) should be used for reporting adverse events occurring in DCTD-sponsored trials.

For DCTD-sponsored trials, each protocol document should include a section giving detailed instructions for reporting AEs to the IND sponsor. This should include a description of both expedited (e.g. AdEERS) and routine reporting. Please refer to the "NCI Guidelines for Investigators: Adverse Event Reporting Requirements for DCTD (CTEP and CIP) and DCP INDs and IDEs" (http://ctep.cancer.gov/protocolDevelopment/electronic_applications/docs/aeguidelines.pdf), which provides reporting tables for the protocol document. All AEs that have a reasonable possibility of having been caused by the agent and that are both serious and unexpected are reported to the FDA in an IND safety report by CTEP when CTEP holds the IND for the study. All other AEs are reported to the FDA in our IND Annual Report.

The expedited reporting of serious AEs is in addition to and does not supplant routine adverse event reporting as part of the regular scientific report of the results of the research protocol. Reporting of adverse events should be in accord with the procedures for reporting results described in Section 10.

12 The Investigator and Protocol/Group Chair: Roles and Responsibilities

12.1 The Investigator

More than twenty thousand investigator-physicians are currently eligible to receive DCTD-sponsored study agents. Most are eligible because they are investigators supported by a peer-reviewed NCI-funded grant, contract, or cooperative agreement. Each investigator agrees to certain essential principles of participation in clinical trials with investigational agents. These principles are contained in an agreement, the FDA Form 1572, http://ctep.cancer.gov/forms/index.html, which is defined by FDA regulation.

Return the completed forms to:
Pharmaceutical Management Branch, CTEP
Division of Cancer Treatment and Diagnosis, NCI
Executive Plaza North, Room 7149
6130 Executive Boulevard, MSC 7422
Bethesda, Maryland 20892-7422

12.1.1 The FDA Form 1572

In signing the FDA Form 1572, the investigator assures CTEP that the clinical trial will be conducted according to ethically and scientifically sound principles. More specifically, a signed FDA Form 1572 commits the investigator to the following obligations or tasks:

- Investigators provide a *statement of the education and experience* on the FDA Form 1572 which qualifies them to perform the study.
- Investigators assure that a properly constituted *IRB will be responsible for the initial and continuing review and approval of the study.* Any changes in the research protocol will require IRB approval, and all unanticipated problems involving risks to human subjects must be reported to the IRB. Such changes must also be approved by CTEP (Section 8.6).
- Investigators are responsible for proper, secure storage of study agents and must *maintain adequate agent accountability records* (Section 15).
- Investigators are required to *prepare and maintain adequate and accurate case histories* designed to record all observations and other data pertinent to each patient (Section 10).
- Investigators must *furnish reports to the CTEP as the investigational agent sponsor* (Section 10). In the case of multicenter studies, the coordinating center and the Protocol Chair are responsible for the generation of these reports. Investigators are responsible for submitting data to the coordinating center.
- Investigators are responsible for promptly reporting AEs according to protocol guidelines, which are based on attribution level, grade and expectedness. Some AEs that are unexpected and severe require reporting through AdEERS within 24 hours. (Refer to Section 11)

CTEP has devised a detailed policy statement that adapts this necessarily broad language to the setting of cancer clinical trials.

- Investigators shall *maintain agent accountability records and case histories for 2 years* following the date an NDA or BLA is approved for that indication or, for at least 2 years after the IND is withdrawn. CTEP will notify investigators when an IND has been withdrawn (see Section 10.3).

- Upon the request of a scientifically trained and properly authorized employee of the DHHS (either FDA or NCI), *investigators will make records available for inspection and copying.*
- Investigators certify that they will personally conduct or supervise the clinical trials. If other physicians administer study agents, they will do so only under the investigators' direct supervision. Each attending physician who meets the DCTD investigator definition is required to be registered with CTEP by submitting a completed FDA Form 1572, Supplemental Investigator Data Form, Financial Disclosure Form, and CV to PMB annually.
- Investigators certify that they will inform subjects or their representatives that agents are being used for investigational purposes and will obtain the written consent of the subjects or their locally authorized representatives.
- Investigators assure CTEP that they will not initiate studies until the IRB has reviewed and approved them.

12.1.2 Responsibilities of an Investigator for Human Subjects Protection

Investigators' responsibilities for IRB review deserve special mention. Each investigator who participates in NCI-sponsored clinical research must have the research approved by an IRB that has an approved Federal Wide Assurance (FWA) issued by OHRP. The DHHS regulations specify the procedures the investigator and his or her institution must follow to protect human subjects; they include IRB composition and function, as well as the basic elements of the informed consent document. All clinical research sponsored by DHHS must be in compliance with these regulations (Title 45, Code of Federal Regulations, Part 46 http://www.access.gpo.gov/nara/cfr/waisidx_01/45cfr46_01.html and Title 21, Parts 50 and 56, http://www.access.gpo.gov/nara/cfr/waisidx_01/21cfr50_01.html, and http://www.access.gpo.gov/nara/cfr/waisidx_01/21cfr56_01.html). Within DHHS, the Office for Human Research Protections (OHRP), www.hhs.gov/ohrp, administers these regulations with each institution. The OHRP negotiates assurances of compliance with HHS regulations for the protection of human subjects. Under an assurance, the IRB is authorized to review and approve research. Each investigator who participates in NCI-sponsored research must conduct the research at an institution with an OHRP-approved assurance.

12.2   The Protocol/Group Chair

12.2.1 Responsibilities

The Protocol Chair or Group Chair designee of a clinical trial assumes certain responsibilities in addition to those of the participating investigator. Specifically, these include:

- Writing the protocol document
- Assuring that necessary approvals are obtained, including those of the IRB, the sponsor (DCTD), and any others for the protocol and subsequent amendments
- Monitoring the study during its execution, which includes:
  - Reviewing each case record to confirm eligibility
  - Reviewing each case record to determine compliance with the protocol
  - Reporting adverse events
  - Determining any necessary changes in the protocol and the informed consent documents and submitting them as protocol amendments to the clinical site and to CTEP

- Monitoring accrual to the study and stopping the study when the requirements of the study design have been fulfilled
  - Reporting study status changes to CTEP (see <u>Section 8.7</u>).
- Analyzing Results: By assessing each case to determine eligibility, evaluability, adverse events, protocol compliance, and outcome (this assessment should be independent of that of the treating investigator)
- Reporting Results to CTEP: Results should be presented in fully analyzed and tabulated form. The Protocol Chair bears the primary responsibility for this task. In Cooperative Group trials, of course, the statistical center is the Group Chair's most important collaborator in fulfillment of reporting requirements.

12.2.2 Who May Serve as a Protocol/Group Chair?

Since Protocol/Group Chairs are responsible for meeting all NCI requirements for IND agent research, as stated in this handbook, Protocol/Group Chairs must be fully qualified investigators. Trainees may not serve as Protocol/Group Chairs.

The participation of physicians who collaborate with major institutions in clinical trials is an important component of the NCI program. The NCI, the clinical Cooperative Groups, CCOP Research Bases and many Cancer Centers recognize these participants' contributions. To assure and maintain the high quality of clinical research conducted by the clinical trials organizations, it is important that they maintain a strong relationship with their affiliate investigators.

To accomplish this objective, clinical site administrative policies and procedures must give participating investigators easy access to accurate and timely information on matters of scientific importance. They must also guide investigators to comply with Federal regulations. The following guidelines can assist clinical sites in formulating specific policies for strengthening the relationship between affiliate investigators consistent with these goals. The content of the following guidelines applies to affiliates of any clinical site.

We recommend that each clinical site develop a formal affiliate policy consistent with these guidelines.

13.1   Affiliate Investigators Definition

An affiliate investigator of a clinical site is a physician who:
Participates in research clinical trials organized by the clinical site, and
Has satisfied all criteria for affiliate membership as defined by the clinical site.

13.2   Requirements of an Affiliate Investigator

All affiliate investigators:
- Must have demonstrated competence in the treatment of cancer patients as defined by the clinical site
- Must have the ability to accrue a minimum number of patients as set by the clinical site
- Must have established a close cooperative professional relationship with the clinical site through regular participation in group meetings and/or educational sessions sponsored by the clinical site
- Must have successfully passed a *probation period* during which time the affiliate investigator has demonstrated:
  - Ability to enter patients on protocol
  - Ability to comply with the protocol
  - Ability to provide accurate and sufficient data to the clinical site
  - Ability to adhere to the procedures and standards of the clinical site and the CTEP
- Must have an approved Federal Wide Assurance (FWA) from OHRP or be listed as a component of an FWA for the protection of human subjects (Section 12.1.2).

13.3   Responsibilities of Affiliate Investigators

The affiliate investigator must adhere to the procedures of the clinical site and CTEP for the conduct of clinical research by:
- Meeting the record keeping policies of the clinical site
- Making certain that each protocol has the full approval of an authorized IRB prior to involvement of human subjects

- Making certain that each patient signs and is given a copy of the IRB-approved consent form. The consent forms should be maintained on file by the affiliate investigator
- Complying with CTEP and FDA policies concerning investigational agents use, which include as a minimum:
    - Annually filing a signed FDA Form 1572, http://ctep.cancer.gov/forms/index.html, a signed Supplemental Investigator Data Form, a signed Financial Disclosure Form, and CV with CTEP
    - Observing DCTD policy and procedures for the proper and secure storage of study agents, including maintaining NCI Agent Accountability Records, http://ctep.cancer.gov/forms/index.html.
    - Agreeing that primary medical records of patients may be audited in accordance with the policies of the clinical site, CTEP, and FDA (Section 16).

13.4   Responsibilities of Clinical Sites for Affiliates

13.4.1 The Cooperative Groups

The Cooperative Groups have the following responsibilities for their participating affiliate investigators and sites:
- Maintaining an accurate and up-to-date roster of all member/affiliate institutions and CCOP/CCOP component investigators
- Informing CTEP of important actions regarding membership status of its member institutions and affiliate investigators
- Reviewing performance of all members/affiliates and CCOP/CCOP components in periodic and timely fashion (including on-site audits)
- Assuring that all clinical trial sites, members and affiliates, are in compliance with CTEP policies, procedures and guidances
- Ensuring that each affiliate institution/investigator has an approved Federal Wide Assurance with the OHRP, http://www.hhs.gov/ohrp/policy/
- Ensuring that full local IRB approval and continuing review has taken place prior to allowing registration of patients on any protocol or continuance of patients on protocols
- For international affiliates, the Cooperative Group is responsible for ensuring that all of the appropriate HHS and international regulatory documents are in place, including necessary approvals by the respective foreign health authorities as well as a Federal Wide Assurance with the OHRP/DHHS, prior to the conduct of the study by the foreign affiliate.
- For international affiliates, the associated Cooperative Group will be responsible for data collection and audits

13.4.2 The Principal Investigator Responsibilities for Affiliate Investigators in Cooperative Group Protocols

The Principal Investigator in Cooperative Groups has the following responsibilities for affiliate investigators:
- Assuring the Cooperative Group that the affiliate member's performance meets the procedures and standards of the clinical site
- Informing the Cooperative Group of important changes in affiliate member relationships

- Providing the affiliate with accurate and timely information on matters of scientific importance
- Communicating to affiliates in a timely manner all policies (and any changes in policy) on the conduct of clinical research.

13.4.3 Cancer Centers

The Cancer Centers have the following responsibilities for their physician members and for their affiliate (see Section 13.1) institutions:

- Maintaining an accurate and up-to-date list of members and affiliates
- Informing CTEP of important actions taken regarding membership status of its members and affiliate investigators
- Conducting periodic and timely review of the performance of all members and affiliate investigators (including audits of affiliate data)
- Developing and implementing a Data Safety Monitoring Plan in accordance with the guidance policy at: http://www.cancer.gov/clinicaltrials/patientsafety/dsm-guidelines/page2#dsmb_role
- Agreeing to site-visit the affiliate investigator in accord with CTEP policies
- Assuring that all members and affiliate investigators are in compliance with CTEP policies and guidelines (http://ctep.cancer.gov/industryCollaborations2/default.htm#guidelines_for_collaborations) and FDA regulations
- Ensuring that each affiliate institution/investigator has registered an appropriate assurance with the OHRP, http://www.hhs.gov/ohrp/policy
- Ensuring that full local IRB approval has been obtained prior to allowing registration of patients on any protocol and on a continuing basis
- Communicating to affiliate investigators all policies and changes in policy on the conduct of clinical research in a timely manner.

CTEP restricts distribution of study agents to sites where physicians are registered as investigators with CTEP. Specifically, *secondary distribution of study agents from registered investigators to unregistered physicians is prohibited.* All patients on clinical trials involving the use of study agents must receive all treatments with these agents from a registered investigator.

The reason for this policy is clear. Physicians who treat patients as part of a clinical trial must have a commitment to the trial's goals and to the clinical research's methodological requirements. They must be experienced in the evaluation of therapeutic results and adverse event manifestations of anticancer therapy.

All investigators participating in trials supported by CTEP must be formally registered with their clinical site (i.e., Cooperative Group or Cancer Center) and with CTEP. Investigators register with CTEP by completing a FDA Form 1572, a Supplemental Investigator Data Form, CV, and a Financial Disclosure Form, http://ctep.cancer.gov/forms/.

The following sections discuss ordering agent, writing patient-specific orders, and administering CTEP-supplied study agents.

14.1 Ordering Agent from PMB

Only those physicians who are registered with the NCI, have an active registration status, and are listed on the title page of a protocol (single or multi-institution trials) or are a member of a Cooperative Group participating on the trial may order agent from PMB once the trial is approved. Physicians may name specific individuals on their Supplemental Data form who may serve as ordering or shipping designees in their stead.

14.2 Writing a Patient-specific Orders

Patient-specific orders for study agents should be written by NCI-registered investigators participating on the specific trial. If other licensed prescribers write orders, the registered investigator who is officially participating on the trial must co-sign the order.
When writing orders for CTEP-supplied study agents, investigators should comply with their local policies and procedures. It is good clinical practice to include the following information (in addition to the information generally needed for any order) on each order for a study agent:
The identifying number of the protocol on which the patient is being treated.
The fact that the appropriate supply of study agent should be used; this is particularly important when the study agent is commercially available, to avoid error.
A reminder for staff to check to protocol to determine the source of the study agent; often, study agents are supplied by more than one source.

14.3 Administration of Study Agents

Registered health professionals, including physicians in training, may administer study agents under the direct supervision of a registered investigator holding a current FDA Form 1572. In such cases, the registered investigator assumes complete responsibility for the use of these agents.

CTEP recognizes that it is convenient for a Cooperative Group member or Cancer Center physician to ask a local physician to administer protocol treatment to a patient who may have traveled long distances to the clinical site for initial consultation. We believe that this approach is ultimately detrimental to the clinical research effort unless the investigator maintains very careful surveillance. If close monitoring is impossible, it seems much more sensible to require that all treatments be administered by registered investigators. Physicians will have to consider carefully whether a patient being evaluated for study will be able to receive each treatment at the hands of a registered physician. We are confident, however, that the benefits of this policy, in terms of both patient safety and integrity of the research data, far outweigh the disadvantages and are in the long term best interests of both patients on clinical trials and the DCTD's agent development program.

If an investigator requests and the PMB approves the use of a local medical doctor (LMD), the LMD must:

- be an active, registered NCI Investigator.
- agree to treat the patient in accordance with the protocol.
- notify the Responsible Investigator of all adverse events and submit all required protocol data to the Responsible Investigator as outlined in the protocol. The LMD will function as an "affiliate investigator."
- have IRB approval or be covered under the Responsible Investigator's IRB.
- maintain drug accountability records.

The Responsible Investigator must:

- be an active registered NCI investigator.
- remain the Principal Investigator for all protocol treatment, patient care, and for all data collection and protocol evaluations for the patient receiving LMD care.
- inform his/her IRB that the LMD will be treating one of his/her protocol patients.
- verify that the patient has tolerated the therapy without excessive toxicity at a stable dose level prior to PMB's approval of LMD.
- remain responsible for all treatment decisions and for the collection and submission of the study data in accordance with the approved protocol.
- inform, the Cooperative Group's Operations Office and seek their approval if the patient is being treated under a "NCI Cooperative Group" protocol.
- inform the PMB when the patient withdraws from the protocol or protocol treatment is discontinued.

The investigator is responsible for the proper and secure physical storage and record keeping of study agents received from CTEP. Specifically, the investigator must:

- Maintain a careful record of the receipt, use and final disposition of all study agents received from CTEP, using the NCI Investigational Agent Accountability Record Form (still referred to as the DARF), http://ctep.cancer.gov/forms/.
- Store the agent in a secure location, accessible to only authorized personnel, preferably in the pharmacy
- Maintain appropriate storage of study agents to ensure their stability and integrity
- Return unused study agents to PMB at study completion or upon notification that an agent is being withdrawn or has expired

The intent of the agent accountability procedures described in this section is to ensure that agents received from DCTD are used only for patients entered onto approved protocols. FDA regulation requires the record keeping described in this section. Investigators are ultimately responsible for the use of study agents shipped in their name. Even if a pharmacist or chemotherapy nurse has the actual task of handling these agents upon receipt, the investigator remains the responsible individual and has agreed to accept this responsibility by signing the FDA 1572, http://ctep.cancer.gov/forms/index.html.

Investigators, clinical trials personnel and interested parties can find a training module that addresses the intricacies of study agent accountability on the CTEP web site, www.ctep.cancer.gov, in PMB's section under "Investigational Drug Handling Slide Show" (http://ctep.cancer.gov/branches/pmb/idh_slideshow.htm). This training module covers ordering, accounting for, and returning study agents. In addition, the PMB section of the web site also includes a "Frequently Asked Questions" section that can help sites deal with common problems.

15.1   Procedures for Agent Accountability and Storage

Investigators or their designees must maintain a NCI Investigational Agent Accountability Record Form (still referred to as the DARF) for every CTEP- supplied agent. A copy of this form may be found at http://ctep.cancer.gov/forms/index.html.

- Store each study agent separately by protocol. If an agent is used for more than one protocol, investigators or their designees should maintain separate physical storage for each protocol. Remember that CTEP provides and accounts for agents on a protocol-by-protocol basis.
- Maintain separate accountability records if
    - an agent is used for more than one protocol; maintain an accountability form for each protocol
    - CTEP supplies multiple agents for a protocol; maintain a an accountability form for each agent
    - A protocol employs different strengths or dosage forms of a particular agent (e.g., an agent with a 1-mg vial and a 5-mg vial would require a different accountability form for the 1-mg vial than for the 5-mg vial); maintain a an accountability form for each strength or dosage form
    - Agents are stored in various places, e.g., main pharmacy, satellite pharmacy, physician's office, or other dispensing areas; maintain a separate accountability form at each location

- Document other transactions (e.g., receipt of agent, returns to the NCI, broken vials, etc.) on the accountability form.
- Refer to individual protocols for ordering and storage information for CTEP-supplied agents. It is important to note that procedures for agent accountability may differ when PMB provides patient-specifically labeled supplies (e.g., the supplies for a double-blind randomized clinical trial). Please refer to the individual protocol or call PMB at (240) 276-6575if questions arise.
- DCTD-supplied study agents may be transferred, within an institution (intra-institutional transfer) from a completed DCTD protocol to another DCTD-approved protocol that employs the same agent, formulation and strength.
  - Complete and fax (301-402-0429) an NCI Investigational Agent Transfer form, http://ctep.cancer.gov/forms/index.html, to the Pharmaceutical Management Branch (PMB) for each agent transfer.
  - Submit transfer forms within 72 hours of the actual transfer.
  - Transfer of DCTD-supplied study agents from active protocols requires prior PMB approval (telephone 301-496-5725). (See PMB Policy and Guideline on the CTEP Home Page.)
- Inter-institutional transfer of DCTD study agents (transfer between institutions) is forbidden unless the PMB specifically pre-approves or authorizes such transfer.

PMB is seeing more subtle differences (yellow vs. brown tablets, micronized powder vs. soft gelatin capsule, investigational vs. commercial label) that might not be readily apparent to you (the site). A site could execute an after-hours emergency transfer without realizing that a subtle difference was a concern, and subsequently, PMB would be unable to approve the transfer. *PMB strongly recommends obtaining approval before transferring any agent.*

15.2   Study Agent Returns

Many investigators are not aware that unused study agents must be returned to the IND sponsor. DCTD, as the study agent sponsor, is responsible for study agent accountability, which includes receipt, distribution, and final disposition of all study agents. Investigators are required to return agents if:

- The study is completed or discontinued
- The agent is expired
- The agent is damaged or unfit for use (e.g., loss of refrigeration)

In situations where a DCTD agent is no longer required for a completed or discontinued protocol, DCTD procedures permit the transfer to another DCTD-sponsored protocol that is using the identical agent, formulation and strength through completion of the NCI Transfer Investigational Agent Form, NIH-2564-1, http://ctep.cancer.gov/forms/index.html, see Section 15.1.

In situations where there is excess inventory or agent that will expire before it can be used, and you have another DCTD protocol(s) using the identical agent, please contact the Pharmaceutical Management Branch (240-276-6575) for assistance in transferring the agent to another DCTD-sponsored study. Otherwise, return the agents as stated in the steps below.

To return study agents to DCTD:
- Package the agents securely to prevent breakage (enclose within a zip-lock bag)

- Complete the Return Drug List Form, NIH-986. Save a copy for your records.
- Send to the NCI Clinical Repository at the address indicated on the Return Drug Form within 90 days of the agent's expiration or closure of the study. Since agents are not re-used upon return, rush delivery and maintenance of labeled storage temperatures is unnecessary.

15.3   Verification of Compliance

Investigators are reminded that auditors will review compliance with procedures to ensure proper agent usage during site visits conducted under the monitoring program. Specifically, auditors will:

- check for appropriate maintenance of the agent accountability system
- spot-check agent accountability records by comparing them with the patients' medical records to verify that the agents were administered to a patient entered in the recorded protocol
- compare actual inventory with accountability form balances

15.4   Handling of Antineoplastic Agents

There has been considerable concern about the potential risk of chronic exposure to low-level concentrations of antineoplastic agents among health care workers routinely handling these agents. The potential mutagenic activity of antineoplastic agents has been examined *in vitro* and *in vivo*. Urinary alkylating and anthracycline agents have shown mutagenic activity in experimental systems, whereas this has not been demonstrated for most of the antimetabolites and vinca alkaloids. Reports indicate that workers who handle antineoplastic agents may absorb them. In addition, some compounds are carcinogenic in animals and are suspected of being so in humans, but only in patients receiving the agent at therapeutic levels.

No clear evidence indicates, however, that chronic exposure to low-level concentrations of antineoplastic agents has been carcinogenic in health-care workers. Nevertheless, it would seem prudent to consider the adoption of certain precautions in the procedures of workers handling these agents.   Several professional organizations have reviewed the data on this subject in an attempt to develop guidelines for safe handling. While there are now several published sets of guidelines, they do not differ significantly.

We have reproduced the *Recommendations for Handling Cytotoxic Agents,* by the National Study Commission on Cytotoxic Exposure in Appendix VII. Please note that these are guidelines and do not have regulatory or legal force. They are included for your consideration and information.

Other pertinent references include:

- ASHP Guidelines on Handling Hazardous Drugs. Available at: http://www.ashp.org/DocLibrary/BestPractices/PrepGdlHazDrugs.aspx

- The CDC NIOSH has a very comprehensive list of recent articles at http://www.cdc.gov/niosh/topics/antineoplastic/pubs.html.

16.1   Introduction

Monitoring is a key component of any clinical trials program. Quality assurance and monitoring are concerned with the execution of a trial, rather than its conception, and with the quality of the data that support the scientific conclusions. NCI's quality assurance program consists of centralized study monitoring augmented by on-site auditing.   http://ctep.cancer.gov/branches/ctmb/default.htm.

Many individual activities are part of quality assurance, and investigators have recognized some of them as vital to the integrity of clinical trials for years. In particular, the quality control of pathology and radiotherapy has been part of the Cooperative Group program for a long time. More recently, investigators have increasingly recognized the importance of verifying the accuracy of other classes of data.

We shall now discuss in more detail the items that form the major focus of the DCTD-sponsored quality assurance effort. Note that the first two classes of concern (protocol compliance and data accuracy) are really central problems in clinical trials methodology. The fact that they are assessed intensively by the on-site audit program should in no way divert attention from their essential importance to the *scientific* content of clinical trials. Additional information may be found at the CTEP Clinical Trials Monitoring Branch web site http://ctep.cancer.gov/branches/ctmb/default.htm.

16.2   Protocol Compliance

16.2.1   Protocol Deviations

For general information see:
http://ctep.cancer.gov/protocolDevelopment/policies_deviations.htm

16.2.1.1 Errors involving a CTEP-supplied investigational agent

The Food and Drug Administration (FDA) requires that deviations from the protocol specific medication administration standards for NCI sponsored trials must be reported to CTEP, as IND sponsor, for evaluation and tracking. The specific information that should be forwarded for NCI evaluation can be found at http://ctep.cancer.gov/branches/pmb/faq.htm. The required information, including a Corrective Action Plan, should be forwarded to: PMBAfterHours@mail.nih.gov.

The following parties must also be notified as soon as the error is discovered:
- Patient
- Local investigator
- Study PI
- Responsible IRB
- Lead Organization (e.g., Cooperative Group, Consortium, Cancer Center)

If there are questions, contact the Pharmaceutical Management Branch (PMB) at (240) 276-6575 or PMBAfterHours@mail.nih.gov.

16.2.2 Cooperative Groups

The groups have recognized the importance of assessing the extent of protocol compliance for many years. One of the first areas to receive attention was the confirmation of diagnosis. Today, all groups have Pathology Committees or Reference Panels for selected studies; central pathology review reduces one important source of variability in trial results. Furthermore, most Cooperative Groups have quality control in radiotherapy, which consists at least of reviews of port films by group radiotherapists. These reviews are best done prospectively, so that errors can be detected in time to alter subsequent treatment. In the Cooperative Groups, the medical oncology committee or the Group Chair reviews case report forms to establish whether dose adjustments have followed protocol guidelines, and whether appropriate study tests have been obtained.

In most Cooperative Groups, the Group Chair also reviews each case to determine eligibility, evaluableness, and validity of response and adverse event assessment. In some cases, the statistical office accomplishes one or more of these tasks. All of these assessments are performed through review of submitted case report forms or data submitted via a remote data capture system.

16.2.3 Cancer Centers

The majority of Cancer Centers have organized procedures to assess protocol compliance centrally and systemically (Section 3, Clinical Trial Sites). The CTEP on-site audit program evaluates protocol compliance as part of its monitoring visits. Indeed, this is a major focus of a monitoring visit to a Cancer Center, along with the administrative review for central data management, protocol development, and data collection.

16.2.4 CTEP Clinical Agent Development Contractors

Protocol compliance is assessed by the Clinical Trials Monitoring Service (CTMS). Phase 1 and phase 2/3 contractors submit raw data to CTMS, which reviews it carefully for extent of compliance with the protocol. Reports of these evaluations are provided to the investigator and to CTEP. Audits are conducted three times per year for the purposes of source data verification and assessment of compliance with protocol and regulatory requirements.

16.3 Data Accuracy

The importance of verifying the accuracy of the basic data elements used in the analysis of study endpoints is obvious. Data accuracy is assessed during on-site audits by comparison of the research record (e.g., flow sheets) with the primary patient record. Response assessment may be evaluated by examination of radiographs or scans, where relevant.

In many of the early on-site audits performed, CTEP was concerned about the absence of formalized procedures at many centers for assessing these important issues internally. Many institutions lacked central registration mechanisms to enroll patients on trials. Centralized systems of data management were often not available. Some institutions lacked clear procedures for certifying the accuracy of research data. Formal procedures for evaluating the accuracy of response assessment, for example by second-party review, were commonly lacking. As institutions have recognized the importance of these tools for the conduct of clinical trials and have brought them "on-line," the quality of data has improved commensurately.

16.4   Procedural Requirements

As an IND sponsor, the DCTD must verify that its investigators adhere to the various procedural requirements. Specific procedural activities checked at the time of the on-site audit are:

- **Informed Consent** - Investigators must be certain that the patient signs the IRB-approved version of the informed consent before protocol-directed therapy begins. Investigators must write the consent form specifically for the protocol, addressing all elements required by Federal regulations.
- **IRB Approval** - Each protocol must have full approval by the IRB named in the assurance for the institution prior to patient entry. There should be IRB review and continuing approval occurring on at least an annual basis. The IRB must also approve substantive protocol amendments.
- **Agent Accountability** - Each investigator must assure that:
  - All DCTD-supplied IND agents are used only for patients on the specific protocol for which the agents were requested and approved by CTEP.
  - Commercial agent is not used for a patient when CTEP is supplying that agent for the protocol.
  - Sites must maintain an NCI Investigational Agent Accountability Record Form (still referred to as the DARF), documenting the disposition of each unit of agent received from CTEP for each protocol. (http://ctep.cancer.gov/forms/index.html)

On-site auditors will also review reporting adverse events and the quality of record keeping with particular reference to the completeness of the source documentation.

Auditors review each of these areas during on-site audits. In addition, many Cooperative Groups and Cancer Centers maintain internal procedures to assure the quality of data on their trials and assure that sites meet regulatory requirements.

16.5   Components of the Quality Assurance Program Implemented by CTEP

Information regarding the Quality Assurance and Monitoring Program may be found at the CTEP web site at:
http://ctep.cancer.gov/branches/ctmb/clinicalTrials/docs/2006_ctmb_guidelines.pdf.

16.5.1 Monitoring

Monitoring includes following the study's overall progress to ensure that
projected accrual goals are met in a timely fashion
excessive accrual is avoided
eligibility and evaluability rates do not fall below acceptable standards
risks of the study do not outweigh benefits.
Poor performance in any of these areas is cause for concern. Because these activities are performed during study execution, they may directly impact the conduct of the trial.

The Cooperative Groups are performing these tasks according to systematic, formalized procedures. For phase 1 studies, the CTMS performs these duties.

Cancer Center studies are monitored by the CTMS with direct oversight from CTMB.

16.5.2 The On-Site Auditing Program: Purpose and Procedures

Purpose of Site Visit Audits

The on-site audit involves the assessment of three components: IRB/Informed Consent Content, Accountability of Investigational Agents and Pharmacy Operations, Review of Patient Cases. The audit process includes:

- Verification of data accuracy by comparing the clinical site's primary medical records with the case report forms for analysis
- Verification of the presence of an IRB-approved consent form signed by the patient prior to the initiation of protocol therapy
- Verification of IRB approval (and at least annual review and re-approval) of each sponsored study as well as amendments
- Verification that procedures for agent accountability comply with federal regulations and follow CTEP procedures, including maintenance of NCI Agent Accountability Records.
- Assessment of compliance with protocol and regulatory requirements

Outline of Audit Procedures

- Trials to be audited are those involving DCTD study agents and selected prevention trials
- All audits will be conducted by persons knowledgeable about clinical trials methodology and the Federal regulations and NCI policies pertinent to clinical trials
- Audits will be randomly timed
- Audits will be conducted at an average rate of once every three years (except CTMS monitored phase 1 trials). High accruing sites may be audited more frequently.

Adaptations of Basic Procedures to Specific Needs

These basic procedures have been adapted to the several types of clinical trials organizations supported by NCI in the following way:

- Cooperative Groups - Each Group will perform its own program of on-site audits, to be conducted by its staff and/or members with direct oversight by CTMB. CTEP or CTMS staff will attend a percentage of audits as observers.
- Clinical Agent Development Contractors - On-site audit visits are made to the phase 0, 1 and select phase 2 grantee three times each year and to the Pediatric consortium once annually. All are site visited by the CTMS. Non-contract studies may be assigned to CTMS monitoring at the discretion of CTEP.
- Others - This category comprises all others performing DCTD IND agent studies, including RO1/PO1/P50 holders (conducting clinical IND agent trials as part of grant-related activity), and new agent studies groups. Audits will be conducted by teams composed of CTMS staff, CTEP staff, and outside physicians, as deemed necessary by CTEP.
- Cancer Centers – Audits will be conducted by teams composed of CTMS staff, CTEP staff and outside physicians. Audits will occur once every three years.

Relationship Between the Content of the Site Visit Audits and the Type of Clinical Trials Organization

Phase of Study	Study Monitoring	Reporting Mechanism	Auditing
Phase 1 & Select Phase 2	CTMS + CTEP	Bi-weekly CRFs to CTMS Monthly CTMS reports downloads to CDS	CTMS Pharm.D./M.D. once per year CTMS/CRA twice per year
Phase 2 & Phase 3 Coop Grp	Cooperative Group + CTEP	CDS quarterly data submissions	Group physicians, CRAs, nurses once every 3 years
Phase 2 & Phase 3 CC-Single Inst.	CTEP	CDS quarterly data submissions	CTEP + CTMS + peer physicians once every 3 years

16.6   Informed Consent and the Monitoring Program

Many have asked about the legality of outside individuals' review of a patient's primary medical record. The answer is straightforward. No Federal law prohibits external review of a patient's medical record. The regulations of informed consent do require, however, that investigators inform patients about "the extent to which confidentiality of records will be maintained." This means that there is no rule against chart review by outsiders, but that the patient must be told what will be done. For this reason, CTEP requires that each informed consent document for trials it sponsors include a statement with the following language, "A qualified representative of FDA and NCI may review my medical records." CTEP may also suggest including a statement that the NCI's Pharmaceutical Collaborator (manufacturer of the agent) may have access to the records. This access may be necessary for the pharmaceutical collaborator to prepare a New Drug Application (NDA) or Biologic License Application (BLA) for an agent.

Also, please note that medical records are protected from inquiries under the Freedom of Information Act (FOIA). Even if the study is performed under Government sponsorship, records on the investigator's premises are not subject to FOIA requests. Furthermore, patient-related records in Government files are protected from FOIA requests by the Privacy Act. As an additional measure of safeguarding, CTEP removes patient names from documents in its possession.

Each clinical site is responsible for adherence to the Health Insurance Portability and Accountability Act (HIPAA). Information regarding HIPAA for Covered Entities may be found at: http://www.hhs.gov/ocr/privacy/hipaa/understanding/coveredentities/index.htm

16.7   Dealing with Problems Identified During On-Site Audits

CTEP and the Cooperative Groups have a full range of options to deal with problems identified at the on-site audit. In a great majority of cases, the measures are intended to be constructive, educational, and corrective rather than punitive. The actions that are taken vary with the individual case.

All reports of on-site audits are sent to CTEP electronically. All reports are assessed by CTEP staff. When major problems are identified by a Cooperative Group audit, auditors convey this information to the Group Chair and CTEP for further action and investigation immediately. After requesting a written clarification, and following review of the case by the Cooperative Group and/or CTEP, CTEP may apply appropriate measures if the original assessment is confirmed. The options for action include:
- Letter of Warning
- Probationary status
- Suspension of patient registration privileges
- Immediate repeat audit
- Removal of access to study agents
- Notification of FDA if investigational agents are involved (FDA may conduct its own investigation)
- Notification of the Office of Research Integrity if scientific misconduct is a possibility (ORI may conduct its own inquiry/investigation)
- Notification of the Office of Human Research Protection (OHRP), http://www.hhs.gov/ohrp/policy, if issues of patient rights, informed consent, or IRB review are involved (OHRP may conduct its own investigation)

CTEP may direct the following actions in instances of suspected data fabrication or falsification or other possible scientific misconduct:

- Replacement of Principal Investigator
- Termination of grant or contract
- Re-analysis or retraction of published results
- Formal ORI investigation
- Debarment of investigator or other staff from future participation in PHS research.

Historically, investigational agents that were given Group C designation by FDA had reproducible activity in one or more specific tumor types. Such an agent altered or was likely to alter the pattern of treatment of the disease and could be safely administered by properly trained physicians without specialized supportive care facilities. This information is included as history only; CTEP no longer employs the Group C mechanism.

17.1 Special Exceptions

Physicians with patients who are refractory to standard measures, who are ineligible for an ongoing research protocol, and who have a cancer diagnosis for which an investigational agent has demonstrated activity may receive the agent from CTEP as a Special Exception to the policy of administering investigational agents only under a research protocol.

Definition

The Special Exception mechanism is the functional equivalent of a single-patient IND but differs from it in that the investigator may obtain investigational agents directly from CTEP, instead of having to obtain an IND from FDA. CTEP provides this mechanism as a service to the oncology community and to cancer patients. CTEP professional and support staff commits substantial effort to maintaining the Special Exception Program.

We expect that patients treated under the Special Exception mechanism are not eligible for established research protocols. Agents available for Special Exception are always in phase 2 or phase 3 trials. CTEP does not grant Special Exceptions for phase 1 agents.

The purpose of the Special Exception mechanism is to make unapproved investigational agents that have a significant activity against specific malignancies available to cancer patients and investigators who otherwise cannot participate in a clinical trial.

Criteria for Approval of a Special Exception Request

Pharmacists of the Pharmaceutical Management Branch and physician staff members of the Investigational Drug Branch review and approve each Special Exception request on a patient-by-patient basis, based on the following considerations:
Is there a research protocol for which the patient is eligible?
Have standard therapies been exhausted?
Is there objective evidence that the investigational agent is active in the disease for which the request is being made?

A review of past experience with Special Exception protocols indicated that patients experienced considerable adverse events with little significant benefit. As a result, CTEP has attempted to improve selection criteria for patients treated under Special Exception. Considerable evidence must attest to the agent's activity for the requested indication. There should be sufficient data available to provide a reasonable expectation that the agent will prolong survival or improve the quality of life in a cohort of similar patients so treated. Reports of low response rates, or responses of brief duration, or anecdotal reports of an occasional response are not sufficient to justify approval.

- Is the agent likely to benefit this patient?

- Even if the agent has been reported to be active in the disease, both the patient's physicians and CTEP physicians must weigh the patient's specific circumstances.
- Please note that the Special Exception mechanism may not be used as a means to obtain agents to treat a series of patients on protocol, or to do pilot work for an intended study. CTEP tracks these requests and will take whatever measures are necessary to discontinue such practices by an investigator. Agents distributed under Special Exception mechanism are investigational and are subject to FDA regulation and CTEP policy.

Requesting a Special Exception Agent

Requests for Special Exceptions may be made in writing or by telephone to the Pharmaceutical Management Branch. Requests must include:

- The patient's age, sex, diagnosis and date of diagnosis
- The patient's previous cancer therapy, and current clinical status
- The intended dose and schedule of the requested agent
- Any proposed concomitant cancer agents or other therapies, and pertinent laboratory data.
- Explanation of why the proposed use of the investigational agent is a better choice than a commercially available agent.

Responsibilities of Physicians Administering Special Exception Agents
See Appendix VIII.

17.2 Treatment Referral Center (TRC)

Purpose

The Treatment Referral Center (TRC) is a means for NCI to provide information to community oncologists about therapeutic options for cancer patients with emphasis on referral to Cooperative Group studies or Cancer Centers. The TRC uses http://clinicaltrials.gov, PDQ (http://www.cancer.gov/clinicaltrials), CTEP information systems databases, data supplied by the NCI-designated Comprehensive or Clinical Cancer Centers, and consultations with CTEP physicians to maintain a referral list of current active research protocols.

The TRC sometimes provides early access to investigational agents for select patient populations. When research indicates investigational treatments are promising in specific diseases (and in specific disease stages, patient populations and/or after a specific amount of prior therapy), the TRC may draft a Treatment Referral Center Protocol. TRC protocols have set criteria to determine patient eligibility while Special Exceptions are assessed on each request:

- TRC Protocols are employed for highly promising agents for a variety of cancers
- TRC Protocols are similar to simple multicenter clinical trials (large one-arm studies with relatively open eligibility and simple objectives)
- TRC Protocols are used to ensure equitable distribution of investigational agents with limited availability
- TRC Protocols typically collect safety and activity data
- TRC Protocols are initially offered to the NCI-designated clinical and comprehensive Cancer Centers

When Cancer Center Investigators determine that commercially available treatment options or active clinical trials are lacking or unavailable for such patient populations, they may consider using an open Treatment Referral Center Protocol as a treatment option.

Method
- Information will generally be provided about treatment options using the following algorithm:
- First priority will be given to suggesting the patient be referred to a major phase 2 or 3 trial (usually Cooperative Group studies).
- If the patient is ineligible for or unable or unwilling to enter a phase 2 or 3 trial, then the physician would be offered the following alternatives:
  - Referral to a participating Cancer Center for evaluation for an investigational protocol. These protocols would offer therapies with potential activity in a specific disease
  - Standard treatment options; e.g., commercially available agents
  - Special Exception agents, if available

Access to the TRC
Physicians at participating Cancer Centers contact the Pharmaceutical Management Branch at (240) 276-6575 to talk with a TRC staff member or to register patients on TRC protocols. Registration procedures and eligibility criteria are provided in the approved TRC protocol. All investigators must register with CTEP and maintain appropriate records (discussed in Section 15 of this handbook.)

More information is available on the CTEP home page (http://ctep.cancer.gov/)

NCI-Cooperative Group-Industry Relationship Guidelines

Information on CTEP Guidelines for collaborations with industry may be found on the web site:

http://ctep.cancer.gov/industryCollaborations2/default.htm#guidelines_for_collaborations

Information is available on:

- NCI Standard Protocol Language for Collaborative Agreements
- CTEP Interaction with Industry
- Model Agreements
- NCI - Cooperative Group - Industry Relationship Guidelines discuss the following ten topics and can be found at http://ctep.cancer.gov/industryCollaborations2/guidelines.htm

  1. Proprietary Agent(s)
  2. Confidentiality
  3. Indemnification / Liability
  4. Intellectual Property and Extramural Inventions
  5. Access to Clinical Data
  6. Procedures Under Which the Collaborator May Contact Cooperative Groups, Member Institutions or Individual Clinical Investigator(s)
  7. Nature and Form of Information Supplied to the Collaborator
  8. Cooperative Group - Collaborator Agreements
  9. Publications
  10. Financial Disclosure

- Intellectual Property Option Policy

Phase 0 Studies

Phase 0 trials are designed primarily to evaluate the pharmacodynamic activity and/or pharmacokinetic properties of selected investigational agents before initiating phase 1 testing, or determining not to continue agent development. (http://clincancerres.aacrjournals.org/cgi/content/abstract/14/12/3675). One of phase 0 trials' major objectives is to interrogate and refine a target or biomarker assay for drug effect in human samples implementing procedures developed and validated in preclinical models (http://clincancerres.aacrjournals.org/cgi/content/full/14/12/3658). Thus, close collaboration between laboratory scientists and clinical investigators is essential to the design and conduct of phase 0 trials. Given the relatively small number of patients and tissue samples, showing a significant drug effect in phase 0 trials requires precise and reproducible assay procedures and innovative statistical methodology. Furthermore, phase 0 trials involving limited exposure of a study agent administered at low doses and/or for a short period allow them to be initiated under the Food and Drug Administration Exploratory IND guidance with less extensive preclinical toxicity data than usually required for first-in-human phase 1 studies under a standard IND (http://www.fda.gov/downloads/Drugs/GuidanceComplianceRegulatoryInformation/Guidances/UCM078933.pdf).

There are several types of phase 0 trials. One type of phase 0 trial is designed primarily to show that the drug affects the target in human tumor and/or surrogate tissue or that a mechanism of action defined in non-clinical models can be observed in humans. A second type of phase 0 trial can be designed to evaluate clinically the properties of two or more structurally similar analogues directed at the same molecular target. Phase 0 trials can also serve to determine a dosing regimen for a molecularly targeted agent or a biomodulator intended for use in combination with other agents. Lastly, phase 0 trials can be designed to develop novel imaging probes or technologies to evaluate the biodistribution, binding characteristics, and target effects of an agent in humans.

Because of the very limited drug exposure, phase 0 trials offer no chance of therapeutic benefit, which can impede subject enrollment, particularly if invasive tumor biopsies are required. Although challenging, the potential barriers to enrollment can be dealt with successfully or minimized by careful attention to the protocol design and informed consent process. In addition, it may be helpful to discuss the proposed trial and obtain input from institutional bioethics staff in the development of the protocol design and consent document. In designing phase 0 trials, it is important to ensure that participation will not adversely affect a subject's eligibility to participate in subsequent therapeutic trials or adversely delay standard therapy. In addition, receiving a drug as part of a phase 0 trial should not prohibit the subject from enrolling in other protocols with that agent or class of agents. Given the non-therapeutic nature and the very limited non-toxic drug exposure, subjects should not be required to wait the standard 4 weeks for "washout" before starting another trial. Shorter washout periods, such as 2 weeks or less, are probably sufficient. In general, phase 0 subjects should not be excluded from participating in subsequent phase 1 studies. Keeping these points in mind when designing protocols can help overcome some of the potential barriers to enrollment.

Phase 0 trials do not replace phase 1 trials conducted to establish affective dose on molecular target, dose-limiting toxicities and define a recommended phase 2 dose.

Nevertheless, data from phase 0 trials supplement subsequent phase 1 studies by allowing them to begin at a higher, potentially more efficacious dose, use a more limited and rationally focused schedule for PK and PD sampling, and apply a qualified PD analytic assay for assessing target modulation and reliable standard operating procedures for human tissue acquisition, handling, and processing. The timelines for data submission and thresholds for expedited adverse event reporting to CTEP may differ from phase 1 trials depending on the particular protocol. Furthermore, the exploratory IND will be withdrawn when the phase 0 trial is completed.

Well-designed and executed phase 0 trials are feasible and have great potential for improving the efficiency and success of subsequent trials, particularly those evaluating molecularly targeted agents. However, the range of resources required for the preclinical and clinical aspects of phase 0 studies, particularly those evaluating target or biomarker effects, is not available at most academic institutions. Because of the non-therapeutic nature of the trials, third party payers are not likely to cover the associated clinical care costs. At minimum, such phase 0 trials require a dedicated PD assay development laboratory and staff who have the necessary expertise in biomarker analytic assay development and validation, as well as the facilities for clinical human tissue PD and PK studies that can be done in real time. Also necessary are a well-organized system for biospecimen procurement and processing and an efficiently integrated and dedicated team of laboratory and clinical investigators with expertise in the conduct of early-phase trials.

More information about phase 0 trials can be found on the DCTD web site
http://dctd.cancer.gov/MajorInitiatives/Sep0507Phase0Workshop/workshop.htm

Policy Statement: The Conduct of Phase 1 and 2 Trials in Children

Introduction

For more than four decades, the NCI has supported evaluations of new agents and new treatment approaches for children with cancer. This support has contributed to the identification of curative treatments for more than 75% of children with cancer and has allowed children with cancer to have access to a broad range of new anticancer agents. However, despite these advances, in excess of 2000 children and adolescents in the U.S. continue to die from cancer each year. The pace of the decline in childhood cancer mortality slowed beginning in the late 1990s and this trend continued into the new decade. Novel treatment strategies and agents are required to identify curative treatments for these patients. An integral component of the NCI research program for children with cancer is the evaluation of new agents in pediatric phase 1 trials. Phase 1 trials are essential in order for children to benefit from recent advances in molecular biology and agent discovery that have led to the development of new classes of molecularly targeted agents.

Phase 1 trials for children differ in several fundamental ways from those performed in adult populations.
- Adult phase 1 trials are usually conducted at single institutions. Because of the relative rarity of cancer in children, pediatric phase 1 trials can rarely be performed efficiently by a single institution, and for this reason the NCI generally supports multi-institutional pediatric phase 1 trials. In multi-institutional phase 1 trials, it is essential that the flow of information between the participating institutions, the Operations/Data Center, the Group Chair, and the NCI be timely and accurate.
- Pediatric phase 1 trials also differ from adult phase 1 trials in the timing of their initiation and in their starting dose. As described in more detail below, it is common practice to begin pediatric phase 1 trials following completion of the initial adult phase 1 experience with an agent, and to begin the pediatric phase 1 trial at approximately 80% of the recommended phase 2 dose in adults.

Additional information about the design and conduct of pediatric phase 1 trials is available in published position papers and review articles [1-3].

Phase 2 trials for children with cancer have the objective of identifying clinically relevant activity for study agents against specific childhood cancers. Like pediatric phase 1 trials, phase 2 trials almost always require multi-institutional collaboration. Most pediatric phase 2 trials are conducted through the Children's Oncology Group (COG). The NCI-supported COG develops and coordinates cancer clinical trials at its more than 200 member institutions, which include cancer centers of all major universities and teaching hospitals throughout the U.S. and Canada, as well as sites in Europe and Australia. COG conducts phase 2 trials for the more common types of cancers occurring in children. A common strategy for pediatric phase 2 trials is to have a single trial evaluate the activity of the study agent against multiple types of childhood cancer, with each different cancer type evaluated individually within its own stratum. Typically pediatric phase 2 trials enroll 20-30 patients per disease stratum, and they utilize standard two-stage designs.

Selection of Institutions for Participation in Phase 1 Trials
Most NCI-sponsored pediatric phase 1 trials are performed by consortia that include 10 to 20 institutions [e.g., the COG Phase 1/Pilot Consortium and the Pediatric Brain Tumor Consortium]. Member are carefully selected based on their:
- experience in developing and participating in early phase trials
- ability to carefully monitor patients treated on phase 1 studies
- capabilities in reporting clinical data in a timely manner to the Operations/Data Center
- resources for collection of specimens for required correlative and pharmacokinetic studies, and
- ability to contribute to the scientific leadership of the Consortium (e.g., pharmacokinetics, pharmacogenetics, and correlative biology).

Institutions must be committed to offering patients participation in phase 1 trials, timely submission of all required data and blood/tissue specimens, and compliance with federal regulations for the protection of research subjects.

Timing of Initiation and Starting Dose for Pediatric Phase 1 Trials
Pediatric phase 1 trials commonly start once the adult recommended phase 2 dose has been established. The starting dose for adult phase 1 trials is often 10% of the dose found to be lethal to 10% of rodents in toxicology studies. This low dose is selected to minimize the risk of severe adverse events among the first humans receiving new agents. Although the initial dose escalations are large in adult phase 1 studies, it is not uncommon in these studies to evaluate ten or more dose levels before dose-limiting toxicity is reached. Completion of studies with a large number of dose levels requires a substantial number of patients. In contrast, the starting dose for pediatric phase 1 studies is usually 80% of the adult recommended phase 2 dose, and the trial escalates in 25-30% increments as successive cohorts of children are accrued to the study. By taking advantage of the adult phase 1 maximum tolerated dose (MTD) to determine the pediatric phase 1 starting dose:
- Pediatric phase 1 trials commonly require fewer than five dose levels and fewer patients to establish the pediatric MTD.
- The occurrence of unanticipated, severe adverse events at the starting dose levels is minimized, as there is considerable adult experience documenting the agent's adverse event profile.
- All children entered onto a phase 1 study receive a dose of the agent that is near the adult phase 2 dose.

This strategy has been successfully employed for over a decade, and in most cases has allowed the efficient determination of a pediatric MTD that is 80% or more of the adult MTD. With this approach, pediatric phase 1 studies can commence at a relatively early time point in the adult development program of an agent without waiting until the adult program's completion.

For those agents that achieve target levels in adults without causing dose-limiting toxicity (DLT), the initial pediatric experience can generally begin at the dose in adults that resulted in the desired biological/clinical effect. In this setting, dose escalation may be limited to a single dose level above the adult recommended phase 2 dose (RP2D). Selection of the pediatric RP2D can be based on the pharmacokinetic profile of the agent in children, selecting a dose level that achieves adequate drug concentrations based on preclinical and adult experience with the agent.

Appendix III

Prioritization of Agents for Phase 1 Evaluation in Children

Hundreds of new agents are currently under evaluation for cancer indications in adults. Only a small fraction of these can be evaluated in children with cancer as a result of the thankfully small number of children eligible for clinical trials evaluating new agents. Because of this increasing imbalance between the number of new agents potentially available for pediatric evaluation and the number that can actually be evaluated, it is essential to prioritize new anticancer agents for testing in children effectively. Data from pediatric preclinical models may provide information that is useful in prioritizing new anticancer agents for testing in children. NCI supports the Pediatric Preclinical Testing Program (PPTP) based on the premise that a systematic approach to developing preclinical data for specific childhood cancers may provide sufficiently reliable data to allow prioritization of truly active agents for pediatric clinical evaluation. Detailed information about the PPTP and its testing procedures is available at the PPTP web site (http://pptp.nchresearch.org/), as are publications and meeting presentations describing PPTP testing results. Agent prioritization decisions can also be based on the genomic characteristics of specific childhood cancers, as genes that are consistently altered by mutation, copy number change, or loss-of-heterozygosity may highlight cellular pathways for therapeutic exploitation.

Protocol Development and Approval

Pediatric phase 1 protocols developed by the NCI-sponsored pediatric consortia should be preceded by a written Letter of Intent (LOI) from the Consortium to the CTEP LOI Coordinator declaring interest in conducting a particular study. The LOI should describe the hypothesis to be investigated, the general design of the contemplated trial, plus relevant information on accrual capabilities to document feasibility. Protocols are to be developed and submitted, and studies are to be conducted, in accordance with the DCTD "Investigator's Handbook". Consortia Operations Centers communicate the results of the NCI's LOI and protocols reviews to their member institutions and to relevant committees. All protocols utilizing NCI-sponsored investigational agents are to be conducted in accordance with the terms of the "Intellectual Property Option to Collaborators", http://ctep.cancer.gov/industryCollaborations2/intellectual_property.htm, and the NCI Standard Protocol Language for Cooperative Research and Development Agreements (CRADAs) and Clinical Trial Agreements (CTAs).

Agent Distribution

For phase 1 trials utilizing study agents distributed by CTEP, agents may be requested by the Principal Investigator (or their authorized designee) at each participating institution. Pharmaceutical Management Branch (PMB) policy requires that agent be shipped directly to the institution at which the patient is to be treated. PMB does not permit the transfer of agents between institutions (unless prior approval from PMB is obtained.) The CTEP assigned protocol number must be used for ordering all CTEP supplied study agents. The responsible investigator at each participating institution must be registered with CTEP, DCTD, through an annual submission of FDA Form 1572, a CV, the Supplemental Investigator Data Form, and the Financial Disclosure Form. If there are several participating investigators at one institution, CTEP supplied agents for the study should be ordered under the name of one lead investigator at that institution. Active CTEP-registered investigators and investigator-designated shipping designees and ordering designees can submit agent requests through the PMB Online Agent Order Processing (OAOP) application.

Study Monitoring

Pediatric consortia conducting phase 1 studies are responsible for assuring accurate and timely knowledge of the progress of each study by:

- Establishing procedures for assigning dose level (for phase 1/dose escalation studies) at the time a new patient is entered, and assuring that the required observation period has elapsed before beginning a higher dose level;
- Documenting and reporting registration, tracking, and attempts to accrue patients who fulfill NIH HHS Guidelines for accrual of women and minority subjects to clinical trials as specified by NIH HHS Guidelines; (http://grants.nih.gov/grants/funding/women_min/women_min.htm)
- Ongoing assessment of case eligibility and evaluability; and ongoing assessment of patient accrual and adherence to defined accrual goals;
- Handling medical review, quality control, and assessment of patient data promptly;
- Reporting treatment related morbidity (adverse events) and measures to ensure communication of this information to all parties rapidly; and
- Conducting interim evaluation and consideration of measures of outcome, as consistent with patient safety and good clinical trials practice for phase 1 and pilot studies.

Data and Safety Monitoring Policies

Each pediatric consortium conducting phase 1 trials must establish a Data and Safety Monitoring Policy in compliance with NIH and NCI guidelines for data monitoring in phase 1 and pilot studies. The policy must be approved by the NCI Program Director. Information concerning NIH policy is available at http://grants.nih.gov/grants/guide/notice-files/not98-084.html with additional description at http://grants.nih.gov/grants/guide/notice-files/NOT-OD-00-038.html. Information concerning essential elements of data and safety monitoring plans for clinical trials funded by the NCI is available on the NCI web site, http://www.cancer.gov/, in the "Conducting Clinical Trials" section.

Adverse Event (AE) Reporting

Each pediatric consortium conducting phase 1 trials is responsible for establishment of a system for assuring timely reporting of all serious and/or unexpected adverse events. For study agents sponsored by the NCI, this involves reporting to the Investigational Drug Branch (IDB), CTEP via the AdEERS system, http://ctep.cancer.gov/reporting/adeers.html, according to CTEP guidelines specified in each protocol. Each of the member institutions of the consortium is responsible for implementing the procedures established for assuring timely reporting of all serious and/or unexpected adverse events.

Site Visit Monitoring

The pediatric consortia conducting phase 1 trials must establish an on-site monitoring program in accordance with the Clinical Trials Monitoring Branch (CTMB, CTEP). For the COG Phase 1 Consortium, this involves on-site audits conducted once every two years of member institutions by the Clinical Trials Monitoring Service (CTMS). The on-site audit program addresses issues of data verification, protocol compliance, compliance with regulatory requirements for the protection of human subjects, and study agent accountability. Any serious problems with data verification or compliance with

Federal regulations must be reported to the Clinical Trials Monitoring Branch immediately. The Operations/Data Center will be responsible for coordinating development of and compliance with corrective programs in response to audits.

Pediatric Exclusivity

Section 111 of the Food and Drug Administration Modernization Act of 1997 (the Modernization Act) created section 505A of the Federal Food, Drug, and Cosmetic Act (the Act) (21 U.S.C. 355a). Section 505A permits certain marketing applications to obtain an additional 6 months of marketing exclusivity (i.e., "pediatric exclusivity") if the applicant, in response to a Written Request from the FDA, files reports of investigations studying the use of the agent in the pediatric population. The pediatric exclusivity provisions were extended until October, 2012 by the *"Best Pharmaceuticals for Children Act of 2007"*. Guidance from the FDA prescribes the general need for pediatric phase 1 studies as a condition for obtaining "pediatric exclusivity" (see http://www.fda.gov/downloads/Drugs/GuidanceComplianceRegulatoryInformation/Guidances/UCM080558.pdf). The guidance recommends that when planning pediatric protocols, pharmaceutical sponsors should discuss protocol designs with a pediatric cooperative study group, as these groups have experience, expertise, and resources that can help applicants optimize their study designs and accrue patients. The NCI-sponsored consortia for conducting phase 1 trials, as well as CTEP staff, are available to assist pharmaceutical sponsors in evaluating whether their agents warrant consideration for pediatric exclusivity, and if so, the design of the early phase studies that would be appropriate to conduct in order to obtain exclusivity.

Reference List

1. Smith M, Bernstein M, Bleyer WA, et al. Conduct of Phase I trials in children with cancer. J Clin Oncol 1998:16(3):966-978.
2. Bernstein ML, Reaman GH, Hirschfeld S. Developmental therapeutics in childhood cancer. A perspective from the Children's Oncology Group and the US Food and Drug Administration. Hematol Oncol Clin North Am 2001:15(4):631-655.
3. Lee DP, Skolnik JM, Adamson PC. Pediatric Phase I trials in oncology: an analysis of study conduct efficiency. J Clin Oncol 2005:23(33):8431-8441.

Key Contact Information

For Overnight Commercial Carrier, use the following street address:	For all other U.S. Mail correspondence, the address is as follows:
National Institutes of Health National Cancer Institute Division of Cancer Treatment and Diagnosis 9609 Medical Center Dr. Room ___ MSC ___ Rockville, MD 20850	National Institutes of Health National Cancer Institute Division of Cancer Treatment and Diagnosis 9609 Medical Center Dr. Room ___ MSC ___ Bethesda, MD 20892
*Please refer to table for appropriate Room / Mail Stop Code (MSC) information.	

CTEP WEBSITE: http://ctep.cancer.gov

Inquiry Type Section/Branch	Room #/MSC	Telephone	FAX	E-mail and/or Website
Protocols and Amendments (Administrative Queries)				
Protocol and Information Office	5W-554 / 9742	(240) 276-6550	(240) 276-7891	PIO@ctep.nci.nih.gov
Clinical Drug Requests				
Pharmaceutical Management Branch	5W-240 / 9725	(240) 276-6575	(240) 276-7893	https://eapps-ctep.nci.nih.gov/OAOP/pages/login.jspx
Drug Information: General Questions and Requests for Non-clinical Use of Investigational Agents				
Agreement Coordination Group	5W-520 / 9740	(240) 276-6580	(240) 276-7894	NciCtepRequests@mail.nih.gov
Drug Information: Pharmacy-related ordering, accountability, preparation information				
Pharmaceutical Management Branch	5W-240 / 9725	(240) 276-6575	(240) 276-7893	PMBafterhours@mail.nih.gov
Investigational Agent Request for Special Exceptions				
Pharmaceutical Management Branch	5W-240 / 9725	(240) 276-6575	(240) 276-7893	PMBafterhours@mail.nih.gov

Inquiry Type Section/Branch	Room #/MSC	Telephone	FAX	E-mail and/or Website
Investigator Brochures (Registered Investigators Only)				
Pharmaceutical Management Branch	5W-240 / 9725	(240) 276-6575	(240) 276-7893	ibcoordinator@mail.nih.gov
Quality Assurance, Site Visits, Monitoring and Informed Consent Issues				
Clinical Trials Monitoring Branch	5W-320 / 9736	(240) 276-6545	(240) 276-7891	smithga@mail.nih.gov
Regulatory Affairs Questions				
Drug Regulatory Group	5W-520 / 9740	(240) 276-6580	(240) 276-7894	casadeij@mail.nih.gov
Clinical Agreement Questions				
Drug Regulatory Group	5W-520 / 9740	(240) 276-6580	(240) 276-7894	NCICTEPACG@mail.nih.gov
Manuscript, Abstract or Press Release Clearance				
Drug Regulatory Group	5W-520 / 9740	(240) 276-6580	(240) 276-7894	NCICTEPpubs@mail.nih.gov
Disease Related Questions (Cooperative Groups and Phase 3 Trials)				
Clinical Investigations Branch	5W-410 / 9737	(240) 276-6560	(240) 276-7892	http://ctep.cancer.gov/branches/cib/default.htm
Agent Related Questions (Early Phase Clinical Trials)				
Investigational Drug Branch	5W-502 / 9739	(240) 276-6565	(240) 276-7894	http://ctep.cancer.gov/branches/idb/default.htm
Possible Scientific Misconduct				
Clinical Trials Monitoring Branch	5W-320 / 9736	(240) 276-6545	(240) 276-7891	smithga@mail.nih.gov
Clinical Grants and Contracts				
Clinical Grants and Contracts Branch	5W-538 / 9741	(240) 276-6540	(240) 276-7891	http://ctep.cancer.gov/branches/cgcb/default.htm
Clinical Trials Operations and Informatics Branch				
Clinical Trials Operations and Informatics Branch	5W-554 / 9742	(240) 276-6550	(240) 276-7891	http://ctep.cancer.gov/branches/pio/default.htm

Informed Consent Resources

Informed Consent Template

The informed consent template is available on the CTEP website:
http://ctep.cancer.gov/protocolDevelopment/default.htm#informed_consent or the NCI
website http://www.cancer.gov/clinicaltrials/patientsafety/simplification-of-informed-
consent-docs/page3.

Appendix 1: Definition of Terms

(http://www.cancer.gov/clinicaltrials/patientsafety/simplification-of-informed-consent-
docs/allpages#appendix1)

Appendix 2: Code of Federal Regulations for the Protection of Human Subjects in Research

(http://www.cancer.gov/clinicaltrials/patientsafety/simplification-of-informed-consent-
docs/allpages#appendix2)

Appendix 3: Checklist for Easy-to-Read Informed Consent Documents

(http://www.cancer.gov/clinicaltrials/patientsafety/simplification-of-informed-consent-
docs/allpages#appendix3)

Appendix 4: Communications Methods

(http://www.cancer.gov/clinicaltrials/patientsafety/simplification-of-informed-consent-
docs/allpages#appendix4)

Appendix 5: Supplemental Materials

(http://www.cancer.gov/clinicaltrials/patientsafety/simplification-of-informed-consent-
docs/allpages#appendix5)

CTEP Glossary

ACTIVATION: The decision by Group/Institution to open a study for patient entry (which occurs only after CTEP approval).

ACTIVATION AMENDMENT: Any protocol *change* that occurs *after* CTEP approval and *prior* to local activation. Examples: CTEP approves the study with recommendations that are incorporated prior to activation; the investigator must list these changes and submit them to CTEP as an activation amendment.

AdEERS: Adverse Event Expedited Reporting System, NCI's web-based system for submitting expedited reports for serious and/or unexpected events forwarded to designated recipients and the NCI for all trials using a NCI-sponsored investigational agent. http://ctep.cancer.gov/protocolDevelopment/electronic_applications/adeers.htm

AE: *Adverse Event* - Any unfavorable and unintended sign (including an abnormal laboratory finding), symptom or disease temporally associated with the use of a medical treatment or procedure regardless of whether it is considered related to the medical treatment or procedure.

AGREEMENT COORDINATION GROUP: a group within the Regulatory Affairs Branch that is responsible for collaborative agreements between NCI and pharmaceutical companies and between NCI and academic institutions

AMENDMENT: *Any* protocol *change* that occurs *after* CTEP approval.

APPROVAL: CTEP approves the protocol in writing when the science and informed consent are acceptable, the IRB documentation is on file (not applicable to Groups), regulatory requirements have been met and the agents to be supplied by the Pharmaceutical Management Branch are available. If recommendations are specified, CTEP expects an "Activation Amendment" to indicate any changes to the approved document.

BLA: *Biologicals License Application* - The formal process by which the FDA makes a biological product generally available to patients and physicians for specific indications.

BRB: Biometric Research Branch, http://dctd.cancer.gov/ProgramPages/brb/default.htm, DCTD, NCI.

CANCER CENTER: *An institution* designated by the NCI as a comprehensive or clinical cancer center and eligible to conduct IND investigational agent studies.

CANCER CLINICAL TRIALS SEARCH: a website to search for information on cancer clinical trials   http://www.cancer.gov/clinicaltrials/search

CAEPR: Comprehensive Adverse Events and Potential Risks list. A list of reported and/or potential adverse events associated with an agent presented by body system.

CCOP: Community Clinical Oncology Program - A cooperative agreement supported program that provides support to community-based oncologists to participate in clinical trials sponsored by clinical cooperative groups and/or cancer centers. Each CCOP is expected to enter a minimum of 50 patients annually on NCI approved research protocols. http://prevention.cancer.gov/programs-resources/programs/ccop

CIB: Clinical Investigations Branch, http://ctep.cancer.gov/branches/cib/default.htm, CTEP, DCTD, NCI.

CIP: Cancer Imaging Program, DCTD, NCI http://imaging.cancer.gov/

CLINICAL TRIAL SITE: An institution, cooperative group, Cancer Center or Consortia that assumes a broad range of responsibilities and functions for the support of clinical trials conducted under its name. It supports the investigator in developing, organizing, implementing, and analyzing clinical trials. It assumes responsibility for the quality of the research, both in concept and execution, and has an important role in assuring patient safety.

CINICALTRIALS.GOV: offers up-to-date information for locating federally and privately supported clinical trials for a wide range of diseases and conditions http://clinicaltrials.gov/

CLINICAL TRIALS MONITORING SERVICE (CTMS): An organization that receives, reviews, and performs data management tasks on individual patient case report forms for phase 1 and some phase 2 NCI investigational agent studies.

CLOSED A: (Protocol Status) Study is closed to accrual, patients still on treatment.

CLOSED B: (Protocol Status) Study is closed to accrual. All patients have completed treatment, but patients are still being followed according to the primary objectives of the study.

COMMERCIAL AGENT: as used in this handbook, a drug (including a biological product) or imaging agent that has been approved by the FDA for commercial distribution for one or more indications, i.e., a lawfully marketed agent. Please refer to the protocol document to determine whether or not a commercially-available agent is being used as an investigational agent in the context of a particular clinical trial. For generic examples of investigational uses of marketed agents, please see the definition of "investigational agent" provided in this glossary. PLEASE NOTE: The combination of an investigational agent with a commercial agent under a CTEP-held IND is considered investigational.

COMPLETE: (Protocol Status) The protocol has been closed to accrual, all patients have completed therapy, and the study has met its primary objectives. A final study report/publication has been submitted to CTEP.

COOPERATIVE GROUPS: Cancer clinical cooperative groups are composed of investigators who join together to develop and implement common protocols. The characteristic of cooperative groups is the central operations and statistical offices which support the administrative requirements of the research and perform central data collection and analysis.

CTCAE: Common Terminology Criteria for Adverse Events. The CTCAE are periodically updated. Please check web site for most recent version. http://ctep.cancer.gov/protocolDevelopment/electronic_applications/ctc.htm

CTEP: Cancer Therapy Evaluation Program, http://ctep.cancer.gov/, DCTD, NCI.

CTMB: Clinical Trials Monitoring Branch, http://ctep.cancer.gov/branches/ctmb/, CTEP, DCTD, NCI.

CTOIB: Clinical Trials Operations and Informatics Branch, http://ctep.cancer.gov/branches/pio/default.htm

CTSU: Clinical Trials Support Unit, https://www.ctsu.org/

DARF- NCI Investigational Agent Accountability Record Form (still referred to as the DARF): Form used to maintain records of disposition of NCI investigational drugs. NIH Form-2564, http://ctep.cancer.gov/forms/default.htm

DCP: Division of Cancer Prevention, http://prevention.cancer.gov/, NCI.

DCTD: Division of Cancer Treatment and Diagnosis, NCI, http://dctd.cancer.gov/

DHHS: Department of Health and Human Services, http://www.dhhs.gov/.

Drug Regulatory Group, Regulatory Affairs Branch (RAB), CTEP, DCTD, NCI. Responsible for CTEP, DCTD INDs and FDA related activities. http://ctep.cancer.gov/branches/rab/default.htm

DTP: Developmental Therapeutics Program, http://dtp.nci.nih.gov/, DCTD, NCI.

FDA: Food and Drug Administration, http://www.fda.gov/, DHHS.

FDA 1572: Also referred to as a "Statement of Investigator;" it is a requirement of Section 505(I) of the Food, Drug and Cosmetic Act and 312.1 of Title 21 CFR, that an investigator complete this form as a condition for receiving and conducting clinical studies involving investigational agent(s). It includes the investigator's training and experience and provides for legal certifications, http://ctep.cancer.gov/forms/index.html.

FEDERAL WIDE ASSURANCE: Under an FWA, an institution commits to HHS that it will comply with the requirements set forth in 45 CFR part 46, as well as the Terms of Assurance. http://www.hhs.gov/ohrp/assurances/assurances/index.html

GROUP CHAIR: The scientific coordinator of the study who is responsible for developing and monitoring the clinical study as well as analyzing, reporting, and publishing its results for a Cooperative Group study.

IDB: Investigational Drug Branch, http://ctep.cancer.gov/branches/idb/default.htm, CTEP, DCTD, NCI.

IND: Investigational New Drug Application - The legal mechanism under which experimental agent research is performed in the United States. An IND is submitted to the Food and Drug Administration in order to receive an exception from premarketing approval requirements so that experimental clinical trials may be conducted.

INVESTIGATOR: Any physician who assumes full responsibility for the treatment and evaluation of patients on research protocols as well as the integrity of the research data.

INVESTIGATOR'S BROCHURE: A confidential document containing all relevant information about the agent, including animal screening, preclinical toxicology, and detailed pharmaceutical data. Also included, if available, is a summary of current knowledge about pharmacology and mechanism of action and a full description of the clinical toxicities.

INVESTIGATIONAL AGENT: a drug (including a biological product) or imaging agent subject to regulation under the Federal Food, Drug, and Cosmetic Act that is intended for administration to humans or animals that has not yet been approved by the FDA and is in the process of being tested for safety and effectiveness in a clinical investigation/trial. This also includes a product with a marketing authorization when used or assembled (formulated or packaged) in a way different from the approved form, when used for an unapproved indication or when used to gain further information about an approved use (Guideline for Good Clinical Practice Section 1.33)

LOI: Letter of Intent-An investigator's declaration of interest in conducting a phase 1, 2, or pilot trial with a specific investigational agent in a particular disease. Approval of the LOI by CTEP commits an investigator to submit a protocol within a specified time frame. http://ctep.cancer.gov/protocolDevelopment/letter_of_intent.htm

NCI: National Cancer Institute, http://www.cancer.gov/, NIH, DHHS.

NDA: New Drug Application - The formal process by which the FDA makes the agent generally available to patients and physicians for specific indications.

NIH: National Institutes of Health, http://www.nih.gov/, DHHS.

OEWG: Operational Efficiency Working Group. Established target timelines and "absolute" deadlines for LOI/Concept review, protocol development, and trial activation (http://ctep.cancer.gov/SpotlightOn/OEWG.htm)

OHRP: Office of Human Research Protection, http://www.hhs.gov/ohrp/

PIO: The Protocol and Information Office is within the Operations and Information Branch, CTEP, DCTD, NCI. PIO manages the protocol and amendment review process, LOIs, and Concepts and maintains the official record of all NCI-sponsored protocols. http://ctep.cancer.gov/branches/pio/default.htm

PMB: Pharmaceutical Management Branch, http://ctep.cancer.gov/branches/pmb/default.htm, CTEP, DCTD, NCI.

PRB: Pharmaceutical Resources Branch, DTP, DCTD, NCI.

PRC: The CTEP Protocol Review Committee reviews and approves all studies involving CTEP-supported study agents

PRINCIPAL INVESTIGATOR (PI): Name of physician who has organizational and fiscal responsibility for the use of federal funds to conduct a clinical study.

PROTOCOL CHAIR: The scientific coordinator of the study who is responsible for developing and monitoring the clinical study as well as analyzing, reporting, and publishing its results for a multi-center (non-Group) trial.

QUALITY ASSURANCE: The monitoring of a clinical trial to assure the quality of the data that supports scientific conclusions.

RAB: Regulatory Affairs Branch, http://ctep.cancer.gov/branches/rab/default.htm, CTEP, DCTD, NCI.

REVISIONS: Any protocol change that occurs prior to protocol approval by CTEP.

RISK PROFILE: An agent-specific list of probable, possible, or definite adverse events attributed to an agent in lay terms for inclusion in the Informed Consent document.

SENIOR CLINICAL INVESTIGATOR: A physician in the IDB who is assigned to an IND agent to coordinate its clinical development. Each investigational agent has a Senior Clinical Investigator assigned to it.

SINGLE PROJECT ASSURANCE (SPA): A formal written agreement with the Office of Human Research Protection (OHRP), (on behalf of the Secretary of DHHS) and an institution which does not have Multiple Project Assurance and conducts a DHHS-sponsored research project. The SPA specifies how the institution will implement the DHHS regulations at 45 CFR 46.

SPEER: Specific Protocol Exceptions to Expedited Reporting. A portion of the CAEPR that lists protocol specific adverse events exceptions to expedited reporting of serious adverse events to the IND sponsor. Previously referred to as the ASAEL.

SPONSOR: An individual, company, institution or organization which takes responsibility for the initiation, management, and/or financing of a clinical trial. (ICH E6) .

TEMPORARILY CLOSED: (Protocol Status) The decision by a Group, Institution, or NCI to stop patient entry pending study evaluation.

National Study Commission on Cytotoxic Exposure: Recommendations for Handling Cytotoxic Agents

Many cytotoxic agents' mutagenic, teratogenic, carcinogenic, and local irritant properties are well established and pose a possible health hazard to occupationally exposed individuals. These potential hazards necessitate special attention to the procedures used in their handling, preparation and administration, and proper disposal of residues and wastes. These recommendations are intended to provide information for the protection of personnel participating in the clinical process of chemotherapy. It is the responsibility of institutional and private health care providers to adopt and use appropriate procedures for protection and safety.

See also: http://www.ors.od.nih.gov/sr/dohs/Pages/default.aspx

Environmental Protection

- Preparation of cytotoxic agents should be performed in a Class II biological safety cabinet located in an area with minimal traffic and air turbulence. Class II Type A cabinets are the minimal requirement. Class II cabinets which are exhausted to the outside are preferred.
- The biological safety cabinet must be certified by qualified personnel at least annually or any time the cabinet is physically moved.

Operator Protection

- Disposable surgical latex gloves are recommended for all procedures involving cytotoxic agents.
- Gloves should routinely be changed approximately every 30 minutes when working steadily with cytotoxic agents. Gloves should be removed immediately after overt contamination.
- Protective barrier garments should be worn for all procedures involving the preparation and disposal of cytotoxic agents. These garments should have a closed front, long sleeves and closed cuff (either elastic or knit).
- Protective garments must not be worn outside the work area.

Techniques and precautions for use in the class II Biological Safety Cabinet

- Special techniques and precautions must be utilized because of the vertical (downward) laminar airflow.
- Clean surfaces of the cabinet using 70% alcohol and a disposable towel before and after preparation. Discard towel into a hazardous chemical waste container.
- Prepare the work surface of the biological safety cabinet by covering it with a plastic-backed absorbent pad. This pad should be changed when the cabinet is cleaned or after a spill.
- The biological safety cabinet should be operated with the blower on, 24 hours per day - seven days a week. Where the biological safety cabinet is utilized infrequently (e.g. 1 or 2 times weekly) it may be turned off after thoroughly cleaning all interior surfaces. Turn on the blower 15 minutes before beginning work in the cabinet.

- Agent preparations must be performed only with the view screen at the recommended access opening. Professionally accepted practices concerning the aseptic preparation of injectable products should be followed.
- All materials needed to complete the procedure should be placed into the biological safety cabinet before beginning work to avoid interruptions of cabinet airflow. Allow a two to three minute period before beginning work for the unit to purge itself of airborne contaminants.
- The proper procedures for use in the biological safety cabinet differ from those used in the horizontal laminar hood because of the nature of the airflow pattern. Clean air descends through the work zone from the top of the cabinet toward the work surface. As it descends, the air is split, with some leaving through the rear perforation and some leaving through the front perforation.
- The least efficient area of the cabinet in terms of product and personnel protection is within three inches of the sides near the front opening, and work should not be performed in these areas
- Entry into and exit from the cabinet should be in a direct manner perpendicular to the face of the cabinet. Rapid movements of the hands in the cabinet and laterally through the protective air barrier should be avoided.

Compounding Procedures and Techniques
- Hands must be washed thoroughly before gloving and after gloves are removed.
- Care must be taken to avoid puncturing of gloves and possible self-inoculation.
- Syringes and I.V sets with Luer-lock fittings should be used whenever possible to avoid spills due to disconnection.
- To minimize aerosolization, vials containing cytotoxic agents should be vented with a hydrophobic filter to equalize internal pressure, or utilize negative pressure technique.
- Before opening ampules, care should be taken to insure that no liquid remains in the tip of the ampule. A sterile disposable sponge should be wrapped around the neck of the ampule to reduce aerosolization. Ampules should be broken in a direction away from the body.
- For sealed vials, final agent measurement should be performed prior to removing the needle from the stopper of the vial and after the pressure has been equalized.
- A closed collection vessel should be available in the biological safety cabinet or the original vial may be used to hold discarded excess agents solutions.
- Cytotoxic agents should be properly labeled to identify the need for caution in handling (e.g., "Chemotherapy: Dispose of properly")
- The final prepared dosage form should be protected from leakage or breakage by being sealed in a transparent plastic container labeled "Do Not Open if Contents Appear to be Broken."

Precautions for Administration
- Disposable surgical latex gloves should be worn during administration of cytotoxic agents. Hands must be washed thoroughly before gloving and after gloves are removed.
- Protective barrier garments may be worn. Such garments should have a closed front, long sleeves and closed cuffs (either elastic or knit)
- Syringes and I.V sets with Luer-lock fittings should be used whenever possible.

- Special care must be taken in priming I.V sets. The distal tip or needle cover must be removed before priming. Priming can be performed into a sterile, alcohol-dampened gauze sponge. Other acceptable methods of priming such as closed receptacles (e.g., evacuated containers) or back-filling of I.V. sets may be utilized. Do not prime sets or syringes into the sink or any open receptacle

Disposal Procedures

- Place contaminated materials in a leak proof, puncture-proof container appropriately marked as hazardous chemical waste. These containers should be suitable to collect bottles, vials, gloves, disposable gowns and other materials used in the preparation and administration of cytotoxic agents.
- Contaminated needles, syringes, sets and tubing should be disposed of intact. In order to prevent aerosolization, needles and syringes should not be clipped.
- Cytotoxic agent waste should be transported according to the institutional procedures for hazardous material.
- There is insufficient information to recommend any preferred method for disposal of cytotoxic agent waste.
    - One acceptable method for disposal of hazardous waste is by incineration in an Environmental Protection Agency (EPA) permitted hazardous waste incinerator.
    - Another acceptable method of disposal is by burial at an EPA permitted hazardous waste site.
    - A licensed hazardous waste disposal company may be consulted for information concerning available methods of disposal in the local area.

Personal Policy Recommendations

- Personnel involved in any aspect of the handling of cytotoxic agents must receive an orientation to the agents, including their known risks, and special training in safe handling procedures.
- Access to the compounding area must be limited to authorized personnel.
- Personnel working with these agents should be supervised regularly to insure compliance with procedures.
- Acute exposures must be documented, and the employee referred for medical examination.
- Personnel should refrain from applying cosmetics in the work area. Cosmetics may provide a source of prolonged exposure if contaminated.
- Eating, drinking, chewing gum, smoking or storing food in areas where cytotoxic agents are handled is prohibited. Each of these can be a source of ingestion if they are accidentally contaminated.

Monitoring Procedures

- Policies and procedures to monitor the equipment and operating techniques of personnel handling cytotoxic agents should be implemented and performed on a regular basis with appropriate documentation. Specific methods of monitoring should be developed to meet the complexities of the function.
- It is recommended that personnel involved in the preparation of cytotoxic agents be given periodic health examinations in accordance with institutional policy.

Procedure for Acute Exposure or Spills
Acute Exposure

- Overtly contaminated gloves or outer garments should be removed immediately.
- Hands must be washed after removing gloves. Some cytotoxic agents have been documented to penetrate gloves.
- In case of skin contact with a cytotoxic agent, the affected area should be washed thoroughly with soap and water. Refer for medical attention as soon as possible.
- For eye exposure, flush affected eye with copious amounts of water, and refer for medical attention immediately.

Spills

- All personnel involved in the clean-up of a spill should wear protective barrier garments (e.g., gloves, gowns, etc.). These garments and other material used in the process should be disposed of properly.
- Double gloving is recommended for cleaning up spills.

Position Statement

Handling of cytotoxic agents by women who are pregnant, attempting to conceive, or breast feeding.

There are substantial data regarding the mutagenic, teratogenic and abortifacient properties of certain cytotoxic agents both in animals and humans who have received therapeutic doses of these agents. Additionally, the scientific literature suggests a possible association of occupational exposure to certain cytotoxic agents during the first trimester of pregnancy with fetal loss, or malformation. These data suggest the need for caution when women who are pregnant, or attempting to conceive, handle cytotoxic agents. Incidentally there is no evidence relating male exposure to cytotoxic agents with adverse fetal outcome.

There are no studies which address the possible risk associated with the occupational exposure to cytotoxic agents and the passage of these agents into breast milk. Nevertheless, it is prudent that women who are breast feeding should exercise caution in handling cytotoxic agents.

If all procedures for safe handling, such as those recommended by the Commission are complied with, the potential for exposure will be minimized.

Personnel should be provided with information to make an individual decision. This information should be provided in written form and it is advisable that a statement of understanding be signed.

It is essential to refer to individual state right-to-know laws to insure compliance.

National Study Commission on Cytotoxic Exposure

Chairman

Louis P. Jeffrey, Sc.D. Director of Pharmacy Services, Rhode Island Hospital Providence, Rhode Island 02902

Commissioners

Roger W. Anderson, M.S. Director of Pharmacy, University of Texas System Cancer Center, M.D. Anderson Hospital and Tumor Institute	Thomas H. Connor, Ph.D. Assistant Professor of Environmental Sciences, Houston School of Public Health, University of Texas Health Sciences Center at Houston	William E. Evans, Pharm.D. Director, Pharmaceutical Division, St. Jude Children's Research Hospital
Clarence L. Fortner, M.S. Head, Drug Management and Authorization Section, IDB,CTEP Division of Cancer Treatment, National Cancer Institute	Joseph E Gallelli, Ph.D. Chief, Pharmacy Department, The Clinical Center, National Institutes of Health	Joseph N. Gallina, Pharm.D. Director of Pharmacy Services, University of Maryland, Medical System Hospital
Dennis M. Hoffman, Pharm.D. Director of Pharmacy Services, University of New York at Stony Brook	Louis A. Leone, M.D. Director of Medical Oncology, Rhode Island Hospital	Suzanne A. Miller, R.N. Oncology Nurse Consultant
Robert M. O'Bryan, M.D. Division Head, Medical Oncology, Henry Ford Hospital		

For additional information contact:
Louis P. Jeffrey, Sc.D., Chairman
National Study Commission on Cytotoxic Exposure
Massachusetts College of Pharmacy and Allied Health Sciences
179 Longwood Avenue Boston. Massachusetts 02115

National Cancer Institute Procedure of Investigational Agents Acquired for Special Exception Treatment of Individual Patients

Food and Drug Administration (FDA) regulations and National Cancer Institute (NCI) policy require the following steps to be completed as indicated:

1) Investigator Registration:
A physician must be registered with the National Cancer Institute as an investigator by having completed a "Statement of Investigator" FDA Form 1572, Supplemental Investigator Data Form (SIDF), Financial Disclosure Form (FDF), and a CV. Forms are available at: http://ctep.cancer.gov/investigatorResources/investigator_registration.htm *If you are NOT currently registered,* a Form 1572, SIDF, and FDF are enclosed, with the understanding you will complete and return these registration forms within 10 working days of their receipt.

2) Protocol:
A brief protocol must be submitted for each patient that describes the treatment plan, toxicity, activity, and monitoring procedures. For your convenience we have devised a standard protocol form which is included and must be completed. Contact the Pharmaceutical Management Branch, Treatment Referral Center at (240) 276-6575to request the form. The original (with signature) must be returned to the Pharmaceutical Management Branch, 6130 Executive Boulevard, Room 7149, Bethesda, MD, 20892, within 10 working days. Please retain a copy for your records.

3) Institutional Review Board Approval:
You must obtain Institutional Review Board Approval *before treating* the patient and retain documentation of this approval in the patient's medical record.

4) Informed Consent:
You must obtain a written informed consent which must be signed by the patient or their guardian *before* treatment. The informed consent must be retained in the patient's medical record. The informed consent should include a reasonable statement about the potential side effects of the agent. The informed consent must address each of the eight elements required under FDA regulations, as detailed on the accompanying sheet.

5) Final Patient Report:
Upon completion of therapy you must provide NCI a report of the treatment experience that describes toxicity and activity. We have enclosed the form, "*The Report of the Independent Investigator.*" Please return this form to the Pharmaceutical Management Branch, Treatment Referral Center Pharmacist, 6130 Executive Boulevard, Room 7149, Bethesda, MD, 20892.

6) Adverse Events:
Reporting of adverse events is required for all NCI Special Exception protocols. The following is a summary of the procedures. For more detailed instructions, computer based training, and the tools used below please see the CTEP home page at: http://ctep.cancer.gov/protocolDevelopment/default.htm#adverse_events_adeers.

Reporting requirements vary according the investigational agent. Please contact PMB for guidance on which AE reporting chart to use in the Special Exception protocol. All reports should be submitted via AdEERS.

Definitions
Adverse Event Expedited Reporting System (AdEERS) – An electronic system for expedited submission of adverse event reports.
http://ctep.cancer.gov/protocolDevelopment/electronic_applications/adeers.htm

Adverse Event - Any unfavorable and unintended sign (including an abnormal laboratory finding), symptom or disease temporally associated with the use of a medical treatment or procedure regardless of whether it is considered related to the medical treatment or procedure.

Attribution – The determination of whether an adverse event is related to a medical treatment or procedure. Attribution categories include Definite, Probable, Possible, Unlikely and Unrelated.

When reporting in AdEERS use the patient's first name and last initial for the patient ID in the patient information section.

Procedure
- Identify the event using the most current version of the Common Terminology Criteria for Adverse Events (CTCAE) (http://ctep.cancer.gov/protocolDevelopment/electronic_applications/ctc.htm).
- Determine the grade or severity of the event by using the CTCAE criteria. The severity is graded between 1 – 5
- Determine Attribution of the event (Definite, Probable, Possible, Unlikely or Unrelated).
- Determine how the event should be reported according to the AE chart embedded in the protocol.

7) Investigational Drug Accountability:
Investigational drug accountability records (Investigational Agent Accountability Record form:
http://ctep.cancer.gov/forms/index.html) must be maintained and retained in your records. These records may be inspected upon request by an authorized representative of the FDA, NCI or agent collaborator.

8) Failure to comply with any of the above procedures may result in suspension of investigator status and prevent further agent shipments on *all* CTEP supported trials.

9) Agent Reorders:
Additional agent may be requested by submitting a request through the PMB Online Agent Order Processing (OAOP) application. You may only order more agent for the patient specifically named on this protocol. The patient's first name and initial of last name should be indicated in the agent request.

Appendix IX

Guidelines for Treatment Regimen Expression and Nomenclature

INTRODUCTION

The Division of Cancer Treatment and Diagnosis (DCTD), National Cancer Institute (NCI), reviews all protocols it sponsors for safety and scientific integrity. Cancer Therapy Evaluation Program (CTEP) staff has developed *Guidelines for Treatment Regimen Expression and Nomenclature* to express chemotherapy regimens in a uniform, clear and consistent manner. The intention is to minimize undue risks to patients on DCTD sponsored investigational clinical trials. CTEP screens all protocol-related documents (e.g., Letters of Intent, Concept Reviews, protocols, protocol amendments, protocol related publications and correspondence) to assure compliance with the treatment regimen guidelines since we cannot approve protocols unless they comply.

Clear, consistent chemotherapy dosage schedules and treatment regimens are important public health issues. Recent events have heightened awareness and concern about the potential for adverse and fatal outcomes as a consequence of medication errors with oncology agents. The American Society of Health-System Pharmacists (ASHP), the American Medical Association (AMA), and the American Nursing Association (ANA) have recommended systematic standardized approaches to reducing medication errors. Their recommendations include educating health care providers and patients regarding appropriate agent therapy, improving collaboration between health care providers, establishing dosage limits, and standardizing a prescribing vocabulary. The *Guidelines for Treatment Regimen Expression and Nomenclature* supplement and reinforce the AMA, ANA, and ASHP recommendations with specific examples illustrating how the guidelines can be applied during protocol development.

CTEP solicited comments and recommendations for the treatment regimen guidelines from clinical pharmacists from comprehensive cancer centers, home infusion services, industry and the Cooperative Group pharmacy chair. Guidelines for expressing dose regimens in treatment plans, agent orders, physician notes and product labeling have also been developed. Investigators should refer to *Standardized Guidelines for Treatment Regimens Expression and Nomenclature*, ASHP 1997, for additional information on this topic.

POLICY

- Instructions for dose regimens should be complete, clear, and simple to follow.
- Treatment regimens should be expressed accurately, completely and consistently throughout a protocol document.

GENERAL GUIDELINES

- **Do not abbreviate** agent names or treatment schedules. Abbreviations can be misinterpreted.

- Use complete approved **generic agent names.** Brand names and abbreviations are not acceptable (*e.g.,* specify 'CARBOplatin' instead of *CBDCA*, 'CISplatin' instead of *CDDP*).
- **Write treatment instructions clearly and explicitly**. No detail (no matter how minor) should be omitted; however, avoid unnecessary redundancy.
- **Delete extraneous information** that may confuse readers (*e.g.*, protocols that use only injectable agents products should not include information for a tablet formulation).
- **Use consistent notation** in expressing quantifiable units, (*e.g.*, either; 1 mcg or 1 mg; qid or Q6h; kg or m^2; either arms or groups)
- **Do not use abbreviations that appear on The Joint Commission/Institute of Medicine "do not use" list.** In particular, do not use trailing zeroes or the Greek letter μ.
    - **Insist that prescribers spell out the word, "units"** out to avoid confusion; a letter "U" can be easily mistaken for a zero and may result in a 10-fold overdose.
    - **Decimal Points** -
        - Never trail a whole number with a decimal point followed by a zero (*i.e.,* "5 mg" not "5.0 mg"). The decimal point may not be seen, resulting in a 10-fold overdose.
        - In expressing units that are less than the whole number one, the dosage should be written with a decimal point preceded by a zero (*i.e.,* "0.125 mg" not ".125 mg"). Without the 'zero' prefix, the decimal point may be missed resulting in a dosing error.
- **Body weight** - Agent dosages may be expressed as a function of body surface area, body weight, or may be calculated to produce a pharmacokinetically-targeted endpoint (*e.g.,* serum or plasma concentration or area under the curve [AUC]).
    - Specify whether clinicians should use absolute (*i.e.,* actual), ideal, or lean body weight to calculate agent dosage as a function of body weight in the treatment plan section.
    - Include the equation describing how clinicians should calculate ideal or lean body weight if you use either.
    - If agent dosage is a function of a calculated pharmacokinetic endpoint, include the equation(s) describing how that value is calculated in the treatment plan.
- **Contiguous treatment days** - Specify the total number of days the agent is administered and the cycle day that treatment commences in the treatment plan. Include parenthetically the cycle days on which treatment occurs.
- **Non-contiguous days** - Specify the cycle days on which each dose should be given in the treatment plan.
- **Cycle (or Course) duration** – Specify the treatment cycle duration (or length). When a treatment regimen is 21 days in duration, the regimen will be repeated on the twenty-second, forty-third, sixty-fourth…, etc. days following treatment initiation.
- **Duration of administration:**
    - Indicate administration duration clearly. If an agent is to be administered on more than one day per cycle, explicitly identify each cycle day.
    - "Day One" typically describes the day on which treatment commences when treatment day enumeration is arbitrary. Avoid using *'day 0 (zero)'*

when describing treatment schedules unless it is necessary (*e.g.,* when describing the day on which hematopoietic progenitor cells are administered after a cytotoxic conditioning regimen in transplantation protocols).

- **Clarify total dose planned per treatment course** - In all treatment plans (protocols) and agent orders, identify and append parenthetically the total dose (as a function of body weight or surface area) that patients are to receive during a treatment course (or cycle).
- **Administration Dates and Times** - When appropriate include specific starting days and times. Be very clear (spell out) in directions for the twelve o'clock hour "12:00 noon" and "12:00 midnight." Expressing time by 24-hour clock notation ('military time') likewise precludes errors due to ambiguous 'a.m.' and 'p.m.' time notations.
- **Treatment information should contain the following elements:**

	Agent Name	Dosage	Administration vehicle name and volume	Administration route	Administration Instructions
Example 1	ABC	200 mg/m^2	0.9% sodium chloride injection 500 ml	Intravenously	Over 1 hour
Example 2	XYZ	50 mg/m^2	NA	Orally	With food

	Administration Schedule	Number of doses to administer, treatment duration, or date when treatment should be discontinued	Starting dates (and times when appropriate)	Total amount of agent administered per course (expressed parenthetically)
Example 1	Every 12 hours	For 6 doses	Start on Day 1	(total dose/cycle = 1,200 mg/m^2)
Example 2	Every morning	For 14 days	Start on Day 1	(total dose/cycle = 700 mg/m^2)

PARENTERAL ADMINISTRATION

Prepare agent products within documented stability and sterility guidelines in accordance with practitioners' local clinical and institutional policies and procedures. Change agent containers at least daily unless extended stability and sterility data are available.

In protocol descriptions and orders for treatment, express agent dosage as the total amount of agent that will be administered from a single agent container, *i.e.,* the total amount of agent per syringe, bag, or other container that will be dispensed.

Exception to this rule: Agent products with extended stability, where an agent is administered from a single container for longer than 24 hours. In such cases, treatment plans and prescribers' orders should specify the amount of agent that is administered during each 24-hour interval. Product container labels should always identify the amount of agent within the container.

For agent admixtures that can be prepared in more than one way, institute reasonable, standard and consistent methods governing how each agent will be prepared and administered.

Include specific fluid volumes and types when possible.

EXAMPLES

Bolus infusion (administration duration ≤ 24 hours):
- Express the amount of agent per container.
- Include the rate of administration, the infusion duration, and days on which the agent is to be administered.

example

"XYZ" 15 mg/m^2 diluted in 50 mL 0.9% sodium chloride injection, infuse intravenously over 15 minutes for one dose on day 1 (total dose/cycle = 15 mg/m^2)

Agent products stable for ≥ 24 hours - (Containers are prepared daily):
- Express the dose per container.
- Include the total dose (as a function of BSA, weight, etc., when appropriate) in parentheses.
- State that the agent must be prepared daily.

example

"XYZ" 8 mg/m^2 per day diluted in 50 mL 0.9% sodium chloride injection, administer by continuous intravenous infusion over 24 hours, daily for three days starting on day 1 (days 1, 2, and 3; total dose/cycle = 24 mg/m^2 over 72 hours). A new IV bag should be prepared daily for 3 days.

Agent products stable for ≥ 24 hours - (Containers are prepared for multiple days):
- Express the dose as the amount of agent administered per day and indicate the number of days for which it is administered.
- Include the total dose (as a function of BSA, weight, etc., when appropriate) in parentheses.
- State that this is a multi-day preparation and for how long the preparation should be infused.

example

"XYZ" 8 mg/m^2 per day diluted in 50 mL 0.9% sodium chloride injection, by continuous intravenous infusion for three days starting on day 1 (total dose = 24 mg/m^2 over 72 hours). This is a multi-day infusion to be infused over 72 hours.

Continuous infusions that require multiple agent product containers:
- Express the dose per container.
- Include the total dose (as a function of BSA, weight…, etc., when appropriate) in parentheses.
- Include the total number of containers used per day.

example

"XYZ" 1 mg/m^2 diluted in 50 mL 0.9% sodium chloride injection, administer by continuous intravenous infusion over three hours, every three hours for three days, starting on day 1 (8 bags/day, total dose = 24 mg/m^2 over 3 days)

ORAL ADMINISTRATION
- Describe agent dosages and schedules as the amount of agent that will be given (or taken) each time the agent is administered, not as a total daily dose that will

be given (or taken) in divided doses, (*e.g.*, 20 mg orally every 6 hours for 5 days vs. 80 mg per day, given in four divided doses for 5 days
- Include guidelines regarding 'rounding-off' doses to the nearest capsule or tablet size.
- Whenever possible, indicate whether agents should be administered (or taken) with food and explain dietary restrictions.

CONCOMITANT (ANCILLARY) MEDICATIONS
- Clearly identified supportive care and essential ancillary medications required by a treatment regimen.
- State complete instructions including appropriate indication, dosage, administration route, schedule, restrictions to use, and any other relevant data explicitly.

TREATMENT MODIFICATIONS
- Define the maximum number of allowable dose reductions before treatment must stop
- include consistent descriptions of modifications among a study's treatment arms for the same agent
- Use consistent terminology for the same meaning (*i.e.,* < grade 3 or < grade 2)
- Describe exactly how a toxicity must resolve before treatment can be resumed or doses re-escalated
- Explain exactly how modifications are to be handled during a cycle or at the start of the next cycle
- Specify how modifying or stopping therapy of one agent impacts the rest of the treatment regimen
- Describe dose modifications as actual doses, e.g. X mg/m^2, and not as a percent of the previous dose
- Use values for CTCAE grades consistent with the actual definition

For dose escalation studies (particularly for patients treated at the initial dose levels), the maximum number of allowable dose level reductions in the dose modification section must be less than or equal to the number of available dose levels defined in the treatment section.

Appendix X

Investigator Handbook Revision History

1) December 2012 Revision 1
Updated Section 4.1.2 Objectives

2) May 2013 Revision 1.1
New Section 3.2.9 Unblinding Procedure for Placebo-Controlled (Blinded) Studies
Updated Appendix IV- Key Contact Information

www.ingramcontent.com/pod-product-compliance
Lightning Source LLC
Chambersburg PA
CBHW051223200326
41519CB00025B/7227

* 9 7 8 1 7 8 2 6 6 5 7 1 7 *